SAVING THE PLANET BY DESIGN

Can we 'save the Planet'? For a resilient, durable and sustainable future for human society, we need to repurpose, reinvent, redesign, remake and recover our human-made world so that our built environment is benignly and seamlessly biointegrated with Nature to function synergistically with it. These are the multiple tasks that humanity must carry out imminently if there is to be a future for human society and all lifeforms and their environments on the Planet. Addressing this is the most compelling question for those whose daily work impacts on Nature, such as architects, engineers, landscape architects, town planners, environmental policy makers, builders and others, but it is also a question that all of humanity needs to urgently address.

Presented here are two key principles as the means to carry out these tasks – 'eco-centricity' being guided by the science of ecology, and 'ecomimesis' as designing and making the built environment, including all artefacts based on the emulation and replication of the 'ecosystem' concept.

Designing with ecology is contended here as the authentic approach to green design from which the next generation of green design will emerge, going beyond current use of accreditation systems. For those who subscribe to this principle, this is articulated here, showing how it can be implemented by design. Adopting these principles is fundamental in our endeavour to save our Planet Earth, and changes profoundly and in entirety the way we design, make, manage and operate our built environment.

KEN YEANG is an architect, planner and ecologist, best known for his signature ecoarchitecture and ecomasterplans, which are differentiated from the work of other green architects by an authentic ecology-based approach, and by their distinctive green aesthetics, performance and biodiversity, beyond conventional rating systems. He was trained at the Architectural Association School in the UK. He is a pioneer in the field of green design, starting with his doctorate in the 1970s at Cambridge University on ecological design and planning.

SAVING THE PLANET BY DESIGN

REINVENTING OUR WORLD THROUGH ECOMIMESIS

KEN YEANG

Routledge
Taylor & Francis Group
LONDON AND NEW YORK

First published 2020
by Routledge
2 Park Square, Milton Park, Abingdon, Oxon OX14 4RN

and by Routledge
52 Vanderbilt Avenue, New York, NY 10017

Routledge is an imprint of the Taylor & Francis Group, an informa business

British Library Cataloguing-in-Publication Data
A catalogue record for this book is available from the British Library

Library of Congress Cataloging-in-Publication Data
Names: Yeang, Ken, 1948- author.
Title: Saving the planet by design : reinventing our world through ecomimesis / Ken Yeang.
Description: New York : Routledge, 2019. | Includes bibliographical references and index.
Identifiers: LCCN 2019020996| ISBN 9780415685832 (hb : alk. paper) | ISBN 9780415685818 (pb : alk. paper) | ISBN 9781315712727 (ebook : alk. paper)
Subjects: LCSH: Sustainable design.
Classification: LCC NK1520 .Y435 2019 | DDC 745.401/12–dc23
LC record available at https://lccn.loc.gov/2019020996

ISBN: 978-0-415-68583-2 (hbk)
ISBN: 978-0-415-68581-8 (pbk)
ISBN: 978-1-315-71272-7 (ebk)

Typeset in Bell Gothic
by Swales & Willis, Exeter, Devon, UK

For Product Safety Concerns and Information please contact our EU representative: GPSR@taylorandfrancis.com
Taylor & Francis Verlag GmbH, Kaufingerstraße 24,
80331 München, Germany.

Contents

ACKNOWLEDGEMENTS

I would like to acknowledge the following:

- Dr Michael John Wells, Ecologist and Ecourbanist, Biodiversity by Design Ltd, for his review of an earlier manuscript
- Cassia Patel, Ecologist, for her review of my manuscript and preparing the references and glossary
- Dr Anuj Jain, Ecologist and conservation biologist, BirdLife International (Asia) Biomimicry Singapore Network, BioSEA and National University, Singapore, for his comments
- Dr James Karl Fischer, Architect and Zoologist, for his comments

Credits:

- Cover design by Eleanor Davis, Cartoonist and Illustrator
- Illustrations by Tuck Leong
- Coordination by Tengku Diyana Hamzah, Assistant Architect, T. R. Hamzah & Yeang Sdn. Bhd.

FOREWORD

I started work on Ecological Design in the 1970s, carrying out research that culminated in my doctoral dissertation at Cambridge University, 'A Theoretical Framework for the Ecological Design and Planning of the Built Environment', which was published as *Designing with Nature: The Ecological Basis for Architectural Design* (McGraw-Hill, 1995). The ideas on the use of the 'ecosystem' concept as an analogy for design were published as 'Bases for Ecosystem Design' (*Architectural Design*, vol. 42, 1972, pp. 434–436) and as 'Bionics – the Use of Biological Analogies in Design' (*Architectural Association Quarterly*, vol. 4, 1974, pp. 48–57), which were further expanded in my *Ecodesign: Instruction Manual* (John Wiley & Sons, 2006, Chapter 1). What followed was a long hiatus when I became occupied with my architectural practice. I resumed research work on this treatise in 2016 concurrent with practice, leading to the completion and publication of this treatise.

PREFACE

This book is written as a set of instructive ideas and principles based on the science of ecology to provide the directives for an ecocentric approach for the making and remaking of our human-made world that includes its built environment and its various production and infrastructure systems. The book advances the earlier work (see below) on the idea of the emulation and replication of the attributes of the 'ecosystem' being ecology's model of Nature as the basis for Ecological Design, this is referred here as 'ecomimicry'. The work's relevance and usefulness are not for just those whose daily work impinges on the natural environment, such as architects, engineers, designers, real estate developers, builders and others, but crucially, for all who are seeking directives for action in enabling a resilient, durable and sustainable future for all of humanity.

My research and theoretical work started as a doctoral dissertation on Ecological Design and planning in 1971. In the intervening years, I had set up and operated a professional architect business in 1976. Implementing the theoretical work into practice required reinventing the design process itself and building prototypes in the interpreting and testing of the theoretical ideas as constructed built form. Practice, while enabling the earlier research and theoretical ideas to be tested, demanded further concurrent investigative research work to be done, and the lessons learnt are summarised in this treatise.

Environmental science makes evidently clear that humanity's endeavours to make and remake our human-made world are now no longer preventive of further environmental degradation, but have become in effect a 'race and rescue' mission. In implementing this mission lies the relevance of this treatise in providing the basis for action in seeking the resilience and sustainability for all lifeforms and their environments on the Planet.

Ken Yeang (Dr)
(kynnet@kyeang.com)
London and Kuala Lumpur

1

Reinventing the human-made world to address the sustainability equation

PROPOSITION

Can we save the Planet by design? – For a resilient, durable and sustainable future for human society, we need to repurpose, reinvent, redesign, remake and recover our human-made world so that our built environment is seamlessly biointegrated with Nature and functions synergistically with it. These are the multiple tasks that we must carry out and must be carried out imminently if there is to be a future for human society and all lifeforms and their environments on the Planet.

Why are these tasks critical? Addressing these is the most compelling question for all those whose daily work impinges on Nature, such as architects, engineers, landscape designers, urban planners, environmental policy makers, builders, real estate developers and others, but it is also a question that all of humanity needs to urgently address. Whereas addressing the issues of environmental impairment had in the immediate past been a question of preventive action to avert further impairments, the progressive state of environmental impairment has advanced such that repairing the existing degradation already done becomes a priority, besides averting further impairments that has now become a 'race and rescue' mission.

Virtually all aspects of the human-made world and its built environment need to be addressed in this way – beginning with humanity's synthetic and semi-synthetic artefacts and systems, including all urban conurbations from mega urban areas and cities to enclosures (such as buildings), as well as engineering structures and infrastructures for urban utilities, including transportation, movement and routing systems, energy production and transfer systems (sometimes referred to as 'grey infrastructure') and all of humanity's artefacts. The list extends to our biotic production systems for generating goods such as food, wood and medicines, among others. Of course, of high priority is our method of energy production from non-renewable sources and industrial systems for the design, fabrication and packaging of the multitude and variety of artefacts that serve both our commercial and our personal needs. The 'remaking' of the above requires that we transition the material flow of systems from the present day's throughput flow of 'take-make-dispose' linear economy to a circular economy of 'take-make-reuse-recycle-replenish-reintegrate (back into the natural environment)'.

How can we carry out the mission outlined above? Our response needs to be based on two key principles presented here – '**ecocentricity**', meaning to work guided by the science of ecology, and '**ecomimesis**', meaning working, designing and making the built environment, including all artefacts, to become Nature-like based on the emulation and replication of the attributes of the '**ecosystem**' as a concept.

What is the significance of these principles? The contention is that adopting these values profoundly changes the way *we design, make, manage and operate our built environment* to achieve our goal of saving and reclaiming the Planet for Nature.

ACHIEVING EFFECTIVE BIOINTEGRATION

The crux in carrying out this multiple set of directives successfully at all instances is effective '**biointegration**' – that is, the seamless and benign connecting and fusing of our human-made world and its technological systems, with its machines and its built environment, with Nature – that is, the natural environment upon which we are completely dependent for resources, from food, water, air and materials to energy. Moreover, this biointegration as reconnecting with Nature must be as seamless and as benign as possible, both physically and systemically.

Biointegration is defined as the state of being combined or the process of combining into a complete and harmonious whole of the physical and systemic connection between the abiotic and the biotic in Nature. An analogy of effective biointegration can be found, for instance, in the medical and dental practice of *orthopaedic* implanting that effectively combines human-made artificial components with the biological. However, we must aspire to do even better, as many of the current orthopaedic implants are made of unsustainable materials. The aim is to create implants that are harmless even after they disintegrate, which must similarly extend by analogy to our built environments, which are in effect implants in Nature, where the materials used must similarly become harmless even after they are dismantled or disintegrated.

This reintegration is crucial, as the current environmental impairment is due to our disconnection from Nature. This broken link needs to be replaced by a synergistic reintegration, which is the complete fusion of the human-made built environment, including not just its artificial but its semi-artificial components, systems and processes, with Nature, and particularly with the Planet's ecosystems, its constituents and its biogeochemical cycles.[1,2,3]

Biointegration can be both physical and systemic. Physical biointegration means minimising and reversing the physical displacement and fragmentation of natural and semi-natural habitats and populations of species by our urban and industrial and infrastructural developments. Systemic biointegration is where the flows and fluxes of energy,[4] materials, water and biota associated with the systems of manufacture, operation and recovery of both artefacts and the built environment are combined and fused together with Nature's ecosystems' and cycles' processes and flows. One of the key aims of effective biointegration must be to avoid the situation whereby we create a mostly inert human-made world and built environment, made from materials taken from various localities over the Planet that are reassembled to be distinctly and physically separated from its surrounding ecosystems and remain and accumulate in this inert state at the end of their used lives.

In effect, if we are able to effectively repurpose and achieve effective biointegration in everything that humans do, build and make in Nature on the Planet, including its built environment and all of its systems, then there will be no environmental problems. Successfully accomplishing this biointegration is then the fundamental challenge in our endeavour to save the Planet.

'SAVING THE PLANET' AS ECOLOGICAL DESIGN

The term 'Ecological Design' here refers to a broad strategic approach that is firstly guided by the science of ecology (see Chapter 2) and secondly affected by the process of 'designing', where the term 'design' is used not as the sole domain of designers, but as an efficient and systematic problem-solving process applicable to all related disciplines, as will be explained later.

What is the purpose of Ecological Design? Ecological Design is pivotal to our mission of saving the Planet. Its purpose and role are to address the issues caused by the millions of artefacts and structures, particularly the extensive urban structures humans design and construct, that make up our 'built environment'. While other species in Nature can make new structures, such as termite mounds and coral reefs, no other species manufactures enduring artefacts from materials that cannot be broken down easily into organic constituents or will only biodegrade over extended timeframes. Many of the artefacts that are for humanity's daily use, together with their packaging, after a short period of use or partial consumption become wastes that are in most instances thrown 'away' rather than being usefully and circularly absorbed back into the environment. In many instances, the packaging of single-use items lasts longer than the contents themselves. It has been estimated that some 80% of the artefacts we produce become problematic waste. The high degree of discarding of unwanted and unrecyclable artefacts and packaging, and the resulting emissions caused by their production, reflect not only the rapid development of human technology and society, but also the profligate and wasteful lifestyle of human society (see Chapter 8).

The waste we emit into the environment includes the voluminous emissions from the production of energy from fossil fuels, waste from the production of food in our animal agriculture and ineffective industrial agricultural systems, wastes from our manufacturing processes, and beyond. Essentially all activities of human society are conducted with total disregard of the consequences of their outputs on the Planet and its natural systems. The Planet, being a 'closed system', is being used as an environmental sink.[5] However, there is currently no other place that becomes the 'away' into which most of humanity's wastes can be 'thrown'. This misuse of the Planet as a sink for humanity's unwanted solids, liquids and gases has reached a level at which this misuse has actually started to change global processes, particularly changing global climate as a result of the gases emanating from the burning of fuels to meet our energy needs, from industrial agricultural and food production from rearing animals, and other urban and semi-urban related activities.[6,7,8] We have similarly begun to affect through our own activities and technologies real changes to other global cycles such as the water cycle, carbon and nitrogen cycles.[9] The scale of these effects has led to the coining of the term 'Anthropocene' to refer to the current geological era in which Planetary processes are being dominated by the actions of humankind.

Human society may not be immediately be aware of the environmental impacts of these on the Planet, as many of the consequences of our actions are not immediately visible to the human eye. Many require a longer timescale to become visibly evident to humanity, and generally even longer to reach the point where humanity can collectively acknowledge that crucial restorative action needs to be imminently taken. This is because the Planet's ecosystems and biogeochemical systems have inherent limits in terms of capacity and speed of process. Being aware of this is a crucial aspect of Ecological Design

Trends are particularly clear in the world's oceans, where temperature increase is evident. Impacts include changing weather conditions, an increase in the intensity and frequency of storms and droughts, and rising sea levels.[10,11] With increasing and constant carbon emissions, Earth's atmosphere traps more heat. As polar ice caps melt, more water enters our oceans, causing rising sea levels. Increased carbon in the atmosphere also means increased carbon in the ocean, causing ocean acidification. Threats appear greatest in coastal areas where so many key human conurbations are located.

When a major climatic phenomenon such as a hurricane or an earthquake creates a 'high-level event' like a storm surge or a tsunami, the effects may be dramatically amplified by combination with background anthropogenic climate change. It is also undeniably evident that humanity is causing a huge acceleration in the loss of biodiversity, both in terms of the loss of tangible components of ecosystems such as habitats and species and in terms of ecosystem processes such as nutrient cycling and **ecological succession** (see Glossary). This is a trend that needs to be halted and reversed. The loss of species has been such as to lend the name of humankind to a mass extinction process – the 'Anthropocene Extinction' – equal in scale and significance to previous mass extinctions caused, for example, by tectonic shifts or meteorite strikes on Earth millions of years ago.

Our impact on the ecosystems is such as to push many ecosystem processes to the point of imminent collapse, threatening humanity's own survival in the process. The devastating events by Nature on human communities are not Nature's retaliation (as implied by some in the popular press). The rest of Nature is indifferent to humanity's motivation. These devastating events are just Nature's physical and chemical reactions to changes that affect the health, resilience and integrity of ecosystems and changes to biogeochemical cycles.[12,13,14,15,16]

Homo sapiens has shown great adaptability through history, adapting quickly to new circumstances when environmental catastrophe strikes and support systems disappear. We move to new areas and find new technological solutions, including substitutes for materials that may end up in short supply or become environmentally problematic. Nature too, given the chance, can be similarly adaptable.[17,18] Species that were in small number or in another area can move in, or whole new species can evolve to adapt to the new conditions. New societies and new Nature can then form new communities, interacting in new processes of succession and adaptation.[19,20]

However, for Nature, the recovery by succession to a new climax community may take hundreds or thousands of years, much longer than human generational timespans. Some species may not be able to keep up with the pace of environmental change. They may find their movements blocked or impeded by human-made barriers such as urban areas and infrastructure, a process referred as 'ecological fragmentation'.[21] In halting and repairing such damage, ecological designers need to work across all scales to connect and protect our irreplaceable landscape across public and private lands, and from cities to the wildest places.

Current Earth sciences indicate that humanity has in effect crossed the critical thresholds of some of the resilience and carrying capacities of the systems operational in the **biosphere** (see Chapter 5 and the Glossary). Human society's impacts on the natural environment have reached the point of no return. This means that even if human society takes imminent, extensive and concerted regenerative action to address and stem the negative consequences of humanity's current actions and to avert future impacts, the effects of humanity's past actions will still exert effects that will persist well into the future.

For example, not all the emitted greenhouse gases can be fully assimilated and sequestered into natural ecosystems, hence an excess remains in the atmosphere.[22] The impacts of these accumulated gases may extend into a future far longer than human history, with consequences for both the environment and humankind.

DESIGNING AND WORKING WITHIN NATURE'S LIMITS OF RESILIENCE

It needs to be emphasised that Ecological Design means working integrally **within the limits of Nature** as functioning within the **carrying capacities** of the Planet's ecological systems and ability to produce natural resources. That means we must respect the limits of resilience and understand the thresholds beyond which ecological perturbation may lead to ecosystem collapse (see page 18).

Natural ecosystems to varying extents have the ability to withstand regular disturbances and disruptions or 'shocks' to their functioning from natural events or the interventions of man. A natural ecosystem generally adjusts to such disturbances so that it continues to support the ecosystem components and processes typically characteristic of it.[23,24] This feature of Nature is often referred to as its natural 'resilience'.

Each organism within an ecosystem, such as a bacterium, protist, fungus, plant or animal (however large or small), plays some role in maintaining the stability of that community and the conditions of its habitat.[25] Theories vary as to how important that role is for each species, and experiments carried out (most at modest geographical scale) have shown direct correlations between biodiversity and the productivity, resilience and stability of different ecosystems. The same experiments indicate that some species do seem to be very much more important than others at a particular time (so-called **keystone species**, see Glossary). However, in a changed environmental condition, that balance can change, and without the wider species diversity, the system as a whole may not always be able to adapt. What is more, when trying to rate the importance of any given species to ecosystem function, it must be borne in mind that at one stage of its life-cycle a particular species may not be particularly important in the normal functioning of a given ecosystem, but at another stage it can be fundamental to that function.

If any of the ecosystem's constituents are diminished or removed by any disruption or disturbance, there may be a degree of substitution whereby a given component of the ecosystem takes up the function previously served by the component now missing. However, ecosystem resilience and adaptive ability are not infinite. With progressive loss of key species, fundamental change to an ecosystem can occur. A threshold is reached at which the ability of the ecosystem to find balance fundamentally changes.

The point at which, with a progressive increase in some environmental perturbation, a threshold of effect is reached at which resilience is exceeded, is generally non-linear. This nonlinearity results from complex web of interactions between ecosystem components and the cascade of secondary effects, for example when a species is lost that has a fundamental effect on nutrient cycling and biomass production. When sufficiently extensive, ecosystemic changes can start to affect biogeochemical cycles and alter atmospheric composition, for example.

After total collapse of an ecosystem, Nature may then recover and 'start over' by the process of ecological succession. However, depending on the extent of devastation, the recovery by succession may happen not over human generational time, but over much longer periods.

A key example of ecosystem limits is the potential ability of forest to absorb excess anthropogenic carbon dioxide emissions. Experiments involving manipulations of carbon dioxide levels have shown that young, rapidly growing forest can increase total net primary production by 25%.[26,27] This response may represent the upper limit for forest carbon sequestration.

A crucial aspect of designing with Nature-based infrastructure (in Chapter 5), then, is the need to work within the critical **limiting factors** of ecosystems. Working within these limiting factors is crucial not only for the ecological designing and making of the built environment, but in influencing the form, content, operation and the life-cycle logistics of all human-made designed systems, artefacts and the material used in them. They also affect approaches to the **recovery** and the eventual **biointegration** of all components of the human-made environment back into Nature (**replenishment**) where achievable. These thresholds further should help determine human efforts in **repairing, restoring and stabilising** the damage already inflicted on the natural environment and the **rejuvenation and regeneration** of the natural environment.

Working within the thresholds and limiting parameters of resilience and carrying capacities of Nature is a further factor underlying Ecological Design. But sadly, rather than working within the sorts of limits and thresholds just described, what we have been doing instead, and are ever more rapidly exacerbating, is to progressively diminish the ability of the biosphere to provide us with our life support system. We have done this by:

- **Directly destroying, displacing and fragmenting natural habitats** by the interventions we make in creating our urban areas, transport networks and food productions and material extraction and waste disposal systems.
- **Treating much of the biosphere as an environmental 'sink'** for non-biodegradable anthropogenic waste, and hence greatly altering many ecosystems and ecosystem processes.
- **Degrading the library of life** – biodiversity in all its forms from gene to habitat to process – often **irreversibly** through both of the above processes.[28]
- **Unsustainably exploiting non-renewable natural resources and unsustainably modifying hydrogeochemical cycles.**

Humanity may indeed have already **crossed the critical thresholds** of the resilience and carrying capacities of the biosphere in many respects. The Stockholm Resilience Centre has identified nine such biospheric thresholds with the limits of which we must operate to ensure a sustainable future both for humankind and that of all other species:[29]

- Available usable land.
- Global climate stability.
- Global hydrological cycle.
- Global nutrient cycles (nitrogen and phosphorus).
- Ocean chemical balance (especially pH).
- Stratospheric ozone layer integrity.
- Chemical toxic loading of organisms.
- Air quality (especially with respect to particulates).
- Biodiversity.

DESTRUCTION, DISPLACEMENT AND FRAGMENTATION OF NATURAL HABITATS

We might contend that humans, in comparison to other species, are 'good at destroying things', especially when we make things. For example, our existing cities can be regarded as ecosystems that are parasitic and dependent upon their surroundings. They require huge areas of land across the Planet where the existing habitats are cleared to provide the food, energy, materials, products and water to our cities and absorb the waste they produce. This can be measured in terms of Global Ecological Footprint. For example, the global footprint of Greater London is roughly equivalent in global area to all the productive land of the UK.

Urban development and industrial-scale agriculture involve extensive clearing of land that leads to destruction of ecosystems, removal of habitats and topsoil, changes to the locality's hydrology, fragmentation of habitats and other destructive impacts.[30] The most stark example of anthropogenic habitat loss relates to our forest and woodland ecosystems. These comprise the predominant natural land cover of most of the Earth's land surface. Tropical forests alone generate about half the terrestrial net primary production on Earth and are known to be the most biodiverse terrestrial ecosystems.[31] However, biodiversity and the effective ability of Nature to assimilate and sequester gaseous emissions from humanity's activities have been greatly reduced by humanity's extensive clearing of global forest land for urban development, for timber and wood products, agriculture, construction of human-made structures such as pipelines, and beyond.

While we cannot fully measure the devastation humanity has caused the natural environment for lack of truly baseline records, more than 50% of the world's forests have been destroyed or degraded by human activity. The amount of old growth forest continues to decline, transformed for agricultural production (especially highly unsustainable cattle ranching), urban development, mining and other uses. Every year, 32 million acres of forest are destroyed. There are notable areas of recovering habitat both through plantation and natural regeneration; however, it will take centuries for cleared and newly regenerating areas to regain the species diversity and ecosystem health of old growth forests.

Human activity also causes destruction by dividing large areas of habitat into smaller, separated patches and regions by breaking up landscapes with constructed artefacts and corridors for transportation. Among the many negative effects of this fragmentation is the isolation of populations, which reduces genetic diversity, exposing species to higher vulnerability to extinction by disease or susceptibility to external disturbance. Smaller patches of habitat also have increased surface area to 'edge effects' that ecosystems fringing human society face, including pollution, emissions and disease.

Additionally, fragmented habitat can be devastating for migrating species that need to travel great distances seasonally for both survival and reproduction. For example, glazed buildings (especially when tall and brightly lit at night) kill millions of birds annually across the globe, especially long-distance migrants, many of which fly at night and at similar elevations. Excavations we create can be pitfall traps for many types of fauna. Transportation routes can serve as deadly barriers. The list goes on.

MISUSE OF THE PLANET AS AN ENVIRONMENTAL SINK

Humanity also uses Nature as an 'environmental sink' for all of society's manufactured outputs and discharges, as mentioned briefly earlier. A key 'service' provided by ecosystems, then, is the sequestering, absorption, uptake and assimilation of the solid, liquid and gaseous waste emissions from all human activities and systems.

Wastes include greenhouse gas emissions (an unfortunate majority of which are discharged to the atmosphere with little treatment), liquid runoff such as human sewage and toxic pollution, as well as many types of discarded and unwanted solid waste that are not readily recyclable or compostable. Adverse impacts of these emissions on the biosphere include climate change and toxic contamination of food chains, which ultimately also impacts human health. Existing cities are the heart of the problem, as their manufacture and operation currently produce 67% of the world's anthropogenic greenhouse gases (projected to rise to 74% by 2030).[32]

While some limited waste outputs can be assimilated and broken down by healthy ecosystems, other wastes may not be, or not as readily. Natural ecosystems have previously encountered some human-made chemical products, for example, over their evolutionary histories. These may include wastes from such activities as mining and many from industry that have concentrated heavy metals, synthesised chemicals and other substances, all of which require specially tailored solutions for disposal at the end of their practical lifespans.

Climatic change caused by anthropogenic emissions of greenhouse gases can trigger a series of secondary effects in the biosphere, such as changes in species distribution, nutrient regimes and resource availability, which can flip ecosystems into new states, states that no longer furnish the ecosystem services they once did. For example, conditions for crop growth are changing over large areas of the globe, and it may not be possible for agricultural systems to adapt to such changes quickly enough to avoid major impact.

Another key concern in terms of global limits is the **stratospheric ozone layer** (see Glossary). This zone of what is a highly dangerous and oxidative gas when at low level in the biosphere (created, for example, by the reaction of nitrogen gases and hydrocarbons from combustion engine exhausts in the presence of sunlight) is by contrast vital to life on Earth as a stratospheric gas layer. It intercepts cancer-inducing ultraviolet (B) rays of the sun. Scientists in 1985 discovered that human-made chemicals called chlorofluorocarbons (CFCs) had been reaching the stratosphere, releasing highly reactive chlorine and destroying this layer, creating a continent-sized hole over the Antarctic. One of the greatest environmental achievements of mankind was the Montreal Protocol – a global agreement made in 1987 to phase out the continued use and manufacture of CFCs – and the hole started to close over again (though is still compromised today).

It is clear that many of our local actions have global implications throughout the entire biosphere through impacts on global processes and cycles, entailing the degradation of ecosystems and communities of species worldwide. We need to tackle the problem at source rather than seeking to apply some elaborate, costly and potentially unsustainable technological fix to address a waste problem that we have created unnecessarily.

Some of the 'pollution' we produce may not seem harmful in public perception. For example, the light pollution emitted by our artificial lighting systems can have very significant adverse effects on behavioural patterns and resulting survival of many species, including insects, birds, bats and turtles. Noise pollution from our activities can also cause harmful effects, disturbing migration, behaviour and communication, particularly in the ocean. Marine noise pollution has caused mass strandings of whales and interferes with the ability of marine mammals to reproduce and survive.

DEGRADING NATURE'S 'LIBRARY OF LIFE' IRREVERSIBLY

Outside the realm of science fiction (in Hollywood), as yet humankind has not found a way to reconstitute lost species by artificial means. DNA has been extracted from fossils, but that is a far cry from reconstituting a lost species. For now, when a species is lost, it is lost forever. When we cause such a loss, our action is **irreversible** and has unforeseen impacts.

The many adaptations that created each species and their characteristics, encompassing the individual and unique chemical processes within each one's physiology, are the product of thousands to millions of years of natural selection. We are inspired by Nature for the manufacture of nearly all engineered and designed products, such as pharmaceuticals, simply because the 'solutions' represented within so many species to different challenges they have faced over their evolutionary history cannot be independently discovered within sensible timescales, if at all, in the laboratory. In essence, the biosphere is the product of a long history of evolution, vital to its current functionality.

We are cutting off the hand that feeds us. Species extinction, habitat degradation and human alteration of ecosystems are an immediate and pressing threat to our ability to feed ourselves sustainably and avoid conflict over resources. While genetic engineering provides us with new options, we remain frighteningly dependent on naturally occurring genetic variation, which is what fuels species adaptation to environmental change.

Any human activity carried out in ignorance of the biodiversity and functioning of existing ecosystems affected by that activity, whether directly or indirectly, risks not only leading to degradation of service provision to mankind, but irreversible losses of the library of life on which we depend for our sustainable future.

We cannot continue to operate in a way that so recklessly endangers the birthright of future generations and their opportunities to 'read through' and benefit from life's library. Moreover, many would argue that this need to change paths should be considered as a **moral duty** to preserve and cherish the rest of Nature of which we form part.

OVER-EXTRACTION OF THE PLANET'S NATURAL RESOURCES

Nature-based infrastructure includes both **renewable and non-renewable material and energy resources**, which serve as the raw materials from which we construct our built environment. Sunlight may be considered an effectively infinite resource, but land and freshwater are certainly not, and we can therefore only produce so much biomass without complete disruption to natural ecosystems.

Humanity needs to preserve, conserve and manage natural resources sustainably for future generations and avoid irreversibly depleting them. This means recycling non-renewables and only consuming renewables at or below the rate that Earth can sustainably reproduce them, taking into consideration the predicted growth in global human population. Looked at in this light, our design systems, whether large or small, and all associated operational processes need to become in effect 'built conservation' – larders of resources – that we can use and reuse.

CONSEQUENCES OF CONTINUED REDUCTION OF NATURE'S PROVISION OF ECOSYSTEM SERVICES

The foregoing should make clear that our current extensive impacts on Earth's ecosystems and the associated enormous loss of global biodiversity are putting increasing pressure on the ability of Nature to provide ecosystem services. Progressive loss of biodiversity will increasingly interfere with ecosystem functionality, hindering Nature's ability to furnish us and other lifeforms with services, which could ultimately lead to the collapse of many ecosystems. At that point, the biosphere will clearly no longer support the current human population of Earth or the populations of many other species. At that point, humanity will have little choice other than attempt to rejuvenate existing ecosystems to restore necessary ecosystem services. At that point, it may be too late for restoration efforts to be effective.

As Nature's provision of these ecosystem services is not evidently visible in public perception, these services tend to be taken for granted by much of humanity. So far, the global human population, including the plethora of social-economic-political systems that it has created, have not exerted a collective demand on leaders, industry or governments to find means to accelerate Nature's ability to provide ecosystem services. This applies especially within the context of the human built environment and all the technological systems within it or associated with it.

Because Nature's provision of ecosystem services is so complex and intricate in components and processes, and interactions operate at a global scale, we cannot yet emulate, substitute or replicate these effectively by technological systems alone, and the civilisation of modern mankind remains reliant on natural and semi-natural ecosystems. We urgently need to give this greater consideration for deriving solutions.

PLACING PRIORITY ON NATURE'S INFRASTRUCTURE AS 'ECOCENTRICITY'

The first step in avoiding the worst depredations of our actions on Nature and to start making reparation for past harm is to ensure that we apply the science of ecology first and foremost in the design process. This is what may be referred to as '**ecocentricity**'. Approaches that start with frameworks or premises that put Nature as secondary to the whims and desires of humankind and human technologies will never result in the long-term survival and benign co-existence of all life on Earth, and are also likely to prevent humankind from achieving sustainable progress. Sooner or later, failure to prioritise Nature-based infrastructure and holistic environmental stewardship will lead to imbalances in natural systems that will in turn destabilise technical, socio-economic or political systems and infrastructures.

WORKING WITHIN NATURE'S THRESHOLDS OF BIOCAPACITY AND RESILIENCE

What are the limitations to Ecological Design? As mentioned earlier, we need to re-examine the limiting parameters of global ecosystems to determine at the outset how to design and operate our human-made world.

In implementing Ecological Design, the designer must be aware of the components of natural infrastructure in every project and apply the precautionary principle, trying to ensure that the ecological integrity of affected ecosystems is maintained or enhanced. In every locality where we are to construct a built system, natural limiting parameters of the ecosystem set limits on the type of designed systems, artefacts and built structures humanity should make, and how these are constructed and produced, used and recovered. They should further influence the material content, mode of manufacture and operation of the built systems over the systems' life-cycles from first use to final resorption into the environment.

All of our endeavours require humanity to work 'with Nature' and to work 'within Nature'. This means that we need to ensure that all of humanity's actions, interventions and product processes, at each step and stage in the logistics of their materials flow in their **life-cycles** (see **life-cycle assessment** in the Glossary), do not exceed Nature's 'levels of tolerances'. These levels of tolerances are Nature's **thresholds** of **resilience and biocapacity** (the 'carrying capacities') that are the crucial parameters of Nature's ecosystems processes and the biosphere's biogeochemical and hydrological cycles (discussed in Chapter 3). Once these thresholds are crossed, irreversible negative ecological consequences to Nature and its systems can ensue.

Ecological thresholds are the points at which slight perturbation in external conditions causes rapid ecosystemic change or changes. When an ecological threshold has been passed, an ecosystem may no longer be able to return to its state of previous balance. Crossing an ecological threshold can then lead to rapid change of what is often referred to as '**ecosystem health**' (see Glossary). Ecological thresholds represent a non-linearity of the responses in ecological or biological systems to external pressures caused by human activities or by natural processes.

The Planet's biogeochemical cycles have thresholds of resilience for the assimilation of the emissions and outputs from the built environment's energy production and manufacturing systems. The existent 'one-way' flow of waste energy and discharges from these production systems needs to be averted, redesigned and remade to 'close the loop' in a benign cyclic, internalised flow within the built environment.

The thresholds of resilience of the biosphere, hydrosphere and geosphere become, then, the limiting parameters within which all of humanity's actions, activities and interventions in Nature must function. We need to move away from the standard urban design where the built environment can be regarded as the 'embodiment of carbon emissions', or perhaps 'environmental damage in physical form'.

Any intervention by humanity that exceeds any aspect of the Planet's biocapacity will not only tend to undermine the Earth's natural systems, but also eliminate or degrade Nature's ability to furnish us with ecosystem services. The ultimate result will be the elimination or degradation of future options for humanity and civilisation.

It should be considered above all that once Nature's thresholds of resilience have been exceeded to a significant extent, humanity's efforts in designing are no longer anticipatory and preventive, but become a question of seeking to 'save the Planet' in an urgent mission or 'rear-guard action' – in other words, a 'race and rescue mission'. The longer humanity waits to work towards an ecologically effective goal, the tighter the window and the less time there is for effective action.

These thresholds are, then, key factors that must **constrain** Ecological Design in all of humanity's endeavours and the construction and functioning of its built environment.

ADOPTING THE SCIENCE OF ECOLOGY AS THE BASIS FOR ACTION: THE PRINCIPLE OF 'ECOCENTRICITY'

What is the basis for our acts and activities? The contention here is that all of our endeavours that impact adversely on the Planet's systems, and most especially our endeavours in designing, making, using and recovering things, need to be fully guided and informed by the **science of ecology** for us to meet our needs while allowing Nature simultaneously to recover and thrive. Ecology must play an undeniable central role at all instances in carrying out the above endeavours.

Ecology is the study of the interactions between the biotic constituents in Nature that include all of its lifeforms and organisms (biotic constituents) and their abiotic physical environments, acting together as whole systems.

Ecological Design is defined as design that is informed by ecology so as to conserve, protect and benefit ecosystems and habitats in the presence of human activities, the human-made world and the built environment. It is an approach that puts Nature at the centre of all design thinking – or in other words, it is an *ecocentric* approach. This approach is referred to here as '**ecocentricity**', a central principle in Ecological Design.

The argument, then, is that design thinking as Ecological Design thinking needs to be applied at all scales, from a global scale of the Planet's biogeochemical cycles and consequences of the use of natural resources to a regional scale of the built environment, down to a micro-scale of molecular biotic and abiotic constituents of ecosystems.

Ecological Design requires considering the ways in which a design responds to the ecosystems and ecological processes of the locality where the design will be made manifest and operate. It asks how the designed systems can minimise and eliminate any negative impacts on habitats and the associated flora and fauna.

It asks, moreover, how design can contribute to biodiversity and the preservation or restoration of habitats, natural capital and ecosystem services. Ecological Design is cognisant of humankind's past and current collective ignoring or denial of the principles of ecology and the limitations of the Planet's natural resources that have led directly to the current state of Planetary environmental degradation. Ecological Design's contention is that the ecological sciences must be the underlying crucial, irrevocable and fundamental basis for our working with Nature. Working in this way is the very definition of Ecological Design.

In a nutshell, ecology must be the vitally crucial discipline that most influences all of humanity's actions, productions and interventions on Earth, the making, remaking and integration of the built environment, and the repairing of the existing dysfunctional relationships and disconnections with Nature and helping it to 'rejuvenate'. At all times in these efforts, we need to ensure that the Planet's ecosystems (see Chapter 3) are healthy, balanced, functional and viable over the long term.

The complex and intricate interconnectivity and interdependence of the Planet's natural systems mean that in many instances the full ecological implications and outcomes of humanity's actions and interventions may not be fully understood or known with certainty. The understanding of ecosystems, of how to avert environmental decline and achieve net gain, is a complex endeavour; where designing and implementing crucially require a multidisciplinary approach to understanding all the intricate processes and interactions in the ecosystems.

REMAKING THE BUILT ENVIRONMENT TO BE IN SYMBIOSIS WITH NATURE

For our built environment to perform and function as a part of Nature and be an extension of Nature, it needs to work in a symbiotic relationship with Nature. '**Symbiosis**' has been defined as a 'close and long-term **biological interaction** between two different biological organisms in different biological organisms in a **mutualistic, commensalistic** or **parasitic** way' (see Glossary).[33,34] The two organisms are human society and all the other lifeforms in Nature. Achieving this mutualistically beneficial symbiotic relationship between humanity and its built environment with the ecosystems, their lifeforms and the biogeochemical cycles in Nature is the goal of Ecological Design.

REMAKING EXISTING CITIES, CONURBATIONS AND URBAN AREAS

What about our existing cities and urban conurbations? Cities are humanity's largest built environments and artefacts, and the most important artefacts human society makes. Cities are regarded by some as humanity's greatest achievements. But are our existing cities and urban conurbations ecologically benign? It is argued here that we need to radically change and remake our existing and new built environment and its systems in their structure, form and processes to become integral with Nature.

Are our cities ecosystems? This is debatable. The contention here is that our existing cities are not ecosystems, and if this aspect is insisted on by many urban ecologists, then we might regard the existing cities as incomplete ecosystems, and in effect as systems parasitic on Nature, as explained below. The model of a city must change from its existent mostly denatured and mostly inert state of being synthetically and overwhelmingly abiotic, and existing physically distinct and separated from Nature in its making and integration with Nature. We need to repair and reconnect our existing cities' and built environments' dysfunctional relationship to Nature.

It is true that Nature exists in the existent urban realm of many cities, but this is only in the form of the heavily manicured and small green patches as remnants of the ecosystem that had been cleared for the urban development that are all located within the city's built matrix. In some instances, Nature is inserted onto and into built structures, such as within atriums, on terraces and roofs, and mounted on walls. These existences of Nature in many instances may have high levels of biodiversity and are in many places today studied by ecologists. However, the fact remains that our urban areas, such as our existing cities, are often currently significantly inert and abiotic systems. The existing relationship between the built and the natural is not systemically integral, but rather a non-systemic and physical juxtaposition. Of course, every bit of green space in a built environment brings benefit, even if the only possible intervention is a few native flowers in a planter box. Those plants contribute to an ecosystem, albeit in an urban context, that can now support, for example, additional butterflies.

If we compare any existent city to wilderness, the differences are self-evident. The existing cities and urban areas of the Planet, especially the major cities of New York, London, Tokyo, Shanghai,

Dubai and others, do not function as an integrated and complementary part of their natural surroundings. For example, these cities take in energy from polluting sources such as from the burning of fossil fuels whereas most terrestrial ecosystems are driven by renewable energy from the sun. Similarly, although there are (increasingly) waste recycling systems in cities and towns, none yet systemically recycles all of its operational waste materials within itself. The vast majority of our urban areas instead produce large quantities of waste and outputs that are discharged into the environment or into accumulated urban piles such as landfills. The existing cities are so unlike the ecosystems in the wilderness, not just visually but also in functionality, that it is debatable whether we can refer to our cities as ecosystems. There are many other aspects of existing urban areas that distinguish their systems from those of Nature, whether natural or semi-natural ecosystems.

For example:

- Most cities are overwhelmingly synthetic and neither self-contained nor self-sufficient. Natural and semi-natural ecological processes are highly modified.
- Naturally occurring ecosystems develop and change over time in a gradual process from the developing stage to later mature stages that is brought about by the progressive replacement of one community by another until, in favourable circumstances, a **climax community** is established that is generally stable until another major environmental perturbation strikes. Natural ecosystems change by this process of succession, where the changes in their form and structure optimise their use of energy, materials and information. This is not the case in our urban areas as a whole, only in components of them.
- Urban areas are parasitically dependent on Nature, extracting resources without offering benefit in return. Far beyond city limits, natural resources supply required goods (energy, materials, food and water) and needed services (air and water purification).

- The sources of materials for the construction, production systems and operation of most urban areas are again generally external to these urban areas, and the voluminous wastes, often generated after a relatively brief period of use, are not recycled as in naturally occurring ecosystems, but rather discarded into landfill, often far from the site of production of the waste.
- In the great majority of naturally occurring ecosystems on Earth (with the exception of some extreme ecosystems fed, for example, through a sulphur-based metabolism, as in the case of deep-sea thermal vents),[35] the source of energy is the sun. Most existing urban areas, by contrast, run on energy supplied by the combustion of non-renewable fossil-based fuels, the resultant gaseous emissions being the key causes of current global climate change as they are a net addition to the carbon cycle. There are also natural systems that contribute notably to greenhouse gas emissions and which are increasing. For example, beavers cause the emission of methane (25 times more potent a greenhouse gas as carbon dioxide) from impounded wetlands that the species creates along water courses. Due to relief of hunting pressure, numbers are increasing and associated emissions now rising. However, this increase is perhaps best viewed as a return to a previous norm of the carbon cycle before anthropogenic perturbations, whereas anthropogenic emissions from fossil fuel combustion and land use are additional to previously existing natural fluxes.
- The source of water supplying most urban areas is not from harvested rainfall, but rather pumped from external sources to the detriment of local ecosystems (for example, abstraction from groundwater aquifers and rivers).[36,37,38]
- Sources of food are in most cases not located within the boundaries of our towns and cities. They are mostly external to the city or town limits, and often at distant locations, transport from which further consumes more of the finite non-renewable fossil fuels and produces emissions which contribute to air pollution and global climate change.

- Most urban areas around the world still produce far too much waste of all kinds and have grossly inadequate provision for its recycling and reuse.

There are also, more importantly, gross disconnections in the flows of material and energy through the built environment which is generally based on throughput and non-cyclic flows of energy, resources and wastes. Unlike natural ecosystems, urban 'ecosystems' only take from the environment rather than simultaneously replenishing it.

For the above reasons, it is arguable whether we can refer to our existing cities and urban conurbations as 'urban ecosystems'. Existing cities are 'incomplete ecosystems', as human-made parasitic ecosystems that are highly impaired and incomplete, where their boundaries are simply artificially drawn as the city's or town's limit.

It can be concluded that existing cities and towns are generally not '**ecologically effective**' (see Page 34) and are unable to effectively address benignly the ecology of the Planet's systems and need to be remade. Changing undesirable conditions in our existing cities and urban conurbations is a massive global undertaking, but must be the key and urgent goal of Ecological Design if we are to have a sustainable future.

REMAKING THE BUILT ENVIRONMENT AS A 'CONSTRUCTED HYBRID LIVING ECOSYSTEM': THE IDEA OF 'ECOMIMESIS'

What can we use as a model for Ecological Design? If the built environment is to effectively biointegrate with Nature, it must become like Nature. It must seek to become like Nature, and more specifically, it must emulate and replicate the characteristics and attributes of '**ecosystems**', where the ecosystem is regarded by some as ecology's 'model of Nature'. The ecosystem is a community of organic (biotic) and inorganic (abiotic) constituents interacting together to create a dynamic interdependent system with various '**trophic levels**' (see Chapter 3).

The approach to the reinventing of the built environment as a 'constructed hybrid living ecosystem' is referred to here as the emulative process of **biomimicry** (see Chapter 3). This is the emulating and replication of the characteristics and attributes of the ecosystem that is referred to here as '**ecomimesis**' (see Chapter 3) in an approach derived from bionics and biomimicry. The term that is proposed here is a development of the concept of biomimicry in which inspiration for design is drawn from natural forms and individual natural processes in our design and manufacture of artefacts. These concepts are further elaborated in Chapter 3. Reinventing of the human built environment as a fully integrated ecosystem is the key eventual goal of Ecological Design.

Essentially, both the existing and the new built environments that we make need to become as faithfully and symmetrically as possible 'constructed ecosystems' that, while designed and effected by humans, must evolve from Nature to become a functioning part of Nature – not separate from it. In other words, the remaking of the built environment into a constructed ecosystem as a living 'hybrid system' refers to remaking it as a combined and predominately biotic system, and not as one with dispersed and fragmented green areas and patches within.

The application of ecomimesis as ecomimicry is the crucial philosophical ('strategic') approach and a crucial pillar in our applying an ecology-based approach to addressing issues of environment. Ecomimicry is discussed in greater detail in Chapter 3.

ADDRESSING THE 'SUSTAINABILITY EQUATION'

What are the crucial key factors that human society needs to address and resolve in seeking to save the Planet? It is clear that our mission to 'save the Planet' is not a simple matter, but a complex endeavour. The contention is that humanity needs to address a series of factors embodied in what may be thought of as a '**sustainability equation'**. This sustainability equation is the overarching contextual framework for all of our endeavours in accomplishing the set of tasks stated above.

The term 'sustainability' suggests at one level the sustenance of something, simply to keep it going over time, referring here to the Planet and its systems and all of its biota. Our effectively addressing and resolving all aspects of this sustainability equation is the key to achieving a resilient, durable and sustainable future for all of humankind, lifeforms and their environments on the Planet. It is important to note that sustainability already exists in Nature – the contention here is that humans simply need to faithfully mimic the natural world, and in particular the 'ecosystem', to reinvent and remake a sustainable human world.

Human society is now primarily an urban species, and it is predicted that by the year 2100 more than 80% of the global human population may reside in urban centres. It is for this reason that the greatly inefficient and harmful designs of nearly all of our existing cities and urban conurbations need to be remade and those of our new cities and urban conurbations need to be rethought and reinvented.

This reinvention is not to seek at the onset immediate innovative solutions to the problem, but starts with identifying the causes of the problem and framing it in a context that enables innovation. The causes are self-evidently defined as the result of humanity's callous abuse of Nature. The problem is summarily framed as an overarching '**sustainability equation**' that consists of addressing simultaneously the following key goals:

- Providing human society and all other lifeforms with the basic and essential needs for living, such as providing basic shelter for protection from the elements, adequate food and nourishment, provision of clean water and sanitation, efficient mobility for work, play and stay without making contaminating emissions, basic well-managed civic services and amenities for its communities and general well-being. Sustainable food provision is crucial on a Planet where its human population is increasing and by some estimates may exceed 8.5 billion by 2030, while at the same time the ability of the Planet to support that ever-increasing human population is actually declining.
- Rectifying the environmental impairments already inflicted, preventing further deleterious impacts that will exacerbate

existing ecological degradation, and rejuvenating/restoring existing impaired environments.

- Addressing society's humanitarian issues, such as poverty, disease, unemployment, human trafficking, society-destroying drug addiction, organised crime, civil conflict and other societal ills. A further humanitarian issue is how to achieve planned control of human population growth to avoid putting such strain on the Earth's resources that unacceptable environmental impairment and social disparity.
- Remaking the human-made world and its urban built environment to be ecologically effective and ecologically benign in its manufacture, operation and disposal processes.

The relevance of the sustainability equation is that so many approaches to sustainability today are incomplete. Even practising environmentalists tend to implement one aspect and ignore the others. Actions to 'solve' this equation must not be incremental, singular or piecemeal.

The sustainability equation frames the key factors that humanity must imminently and inclusively address, but seamlessly and benignly with Nature. All the challenges in the equation need to be addressed concurrently, and not sequentially. The action has to be concerted, collective and large-scale. It needs to be at least at the infrastructural scale, and where possible at the bioregional and global scale, to be effective. The reasons for this should be clear from the foregoing – because of our current role as global Ecological Designers and also because of the ethical duty of stewardship to Nature that comes with this role. These goals for resolution of the sustainability equation translate into the need to implement these imperative directives for human society.

However, the rectification of the environmental impairments that have already been inflicted on the Planet should be given the highest priority, as all the other factors become considerably easier to resolve if we start out with clean air, clean water, clean land and healthy ecosystems.

In summary, this 'sustainability equation' constitutes the key issues that humanity must address in its endeavours to 'save the Planet'. Addressing these in totality is the fundamental premise and goal of Ecological Design. In effect, whether humanity is able to sustain its survival as a species on Planet Earth in the long term depends on the extent to which it is successful in addressing adequately the challenges inherent in the 'equation'.

The key tasks of Ecological Design in outline

What are the multiple tasks in our endeavour to address the factors in the sustainability equation in the making, remaking and use of the human-made world and the built environment referred at the outset here? The multiple tasks of Ecological Design can be defined as follows:

Ecological repurposing

Is it adequate to just focus on the physical and tangible aspects of Ecological Design? We need to change our thinking about Nature and human society's role in Nature. While being cognisant of the goals in our endeavour to 'save the Planet' and to reclaim Nature, effecting these requires a sea change and epochal re-visioning of societal mindsets, perceptions and ideologies with respect to Nature (see Chapter 9). Ecological Design requires a switch in humanity's mindset from a relationship of exploitation to one of benign stewardship, particularly among those in human society whose daily work most impinges on and potentially threatens the natural environment (see Chapter 7).

The ecocentric mindset and ideology is influenced and guided by the science of *ecology*, which focuses on Nature and seeks a benign and restorative relationship with it and always avoids doing it irreversible harm.

It questions, for instance, in what ways humanity's action and our built systems can respond to the ecology of this place and bioregion, how the built environment can become more connected with the ecology of its locality, with its regional ecosystems and the Planet's biogeochemical cycles, how we can minimise and eliminate the negative impacts on biotic life such as birds and other animals, how humanity's action and built systems can contribute to biodiversity and the preservation or restoration of habitats and enhancement of ecosystem services, and how such action can achieve positive net gains for Nature.

Achieving this change in humanity will require a public pedagogy affected by advocacy, persuasion and education. There will need to be significant efforts by society's 'behaviour changers' in asking humanity to think beyond the anthropocentric purposes of our built environment and artefacts to remake the human-made world and the built environment, as much for the rest of biodiversity as for ourselves, and proselytising people to the ecocentric ideology and approach by the design community in implementing exemplary design by leadership and initiative, and ultimately by all of human society's political leaders to show intent and will (see Chapter 8).

As a key to this end goal of 'saving the Planet' and 'reclaiming Nature', humanity needs to redesign, remake and transform the existing human-made world and new built environments where possible in entirety, to become ecologically benign by becoming like Nature, referred here as becoming ecosystem-like, as 'constructed ecosystems' that are *hybrid living systems*, and not in physical and systemic isolation and ecological alienation from Nature.

Humanity needs to change its perception of its role and that of its built environment in Nature so that its ideology and mindset towards Nature move from regarding Nature as an endless source of natural resources to be exploited and to being a steward and partner of Nature (see Chapter 9).

Ecological redesigning

Ecological redesigning means we need to design in a completely different way to the current norm. This is the most important factor for all designers (such as architects, engineers, town planners and industrial designers). For instance, in most conventional design processes, the tendency is to work inversely (but not at all instances) from the 'inside-out', where the designed system's consequences on ecosystems and biospheric cycles are assessed after the designed system's internal configuration is completed, or partially completed. By contrast, in Ecological Design *the ecological environment external to the designed system (besides others) must drive the design's outcome at the outset of the designing process.*

For instance, before commencing design, we need to understand the locality's ecology and ecological services that are available on-site and throughout the bioregion. Design needs to develop the site in a way that protects or restores or enhances these ecological services that link the land with its ecological history. Where ecology becomes a crucial influencing determinant of design, the external environment and everything else change. The focus becomes as much as on human shelter and amenity.

Account must be taken not just of the ecological consequences of the designed system's physical displacement or disturbance of ecosystems, but of the interaction with ecosystems of the multiple materials used in design at each stage of their life-cycle from the time that the raw material is initially extracted from the Planet to eventual disintegration of the artefacts of design and their eventual environmental reintegration, replenishing Nature.

The redesigning in Ecological Design further requires the re-envisioning of the way human society thinks, conceives and gives purpose to the creation of its physical systems (see Chapter 7). The revised approach to redesigning, for instance, requires at the outset our questioning of the very need of the designed system itself.

Ecologically redesigning therefore requires the reformulating of the fundamental premises for design so that they become as indicated earlier, **ecology-centric** – an approach that gives deference and centrality to the health of the Planet.

Adopting an ecology-centric (referred to here as 'ecocentric') design is an environmentally benign approach that averts making negative impacts while concurrently rectifying past deleterious human damage to the environment, and avoiding repetition of such damage. This must be carried out holistically, and not incrementally nor in a piecemeal way, such as, for instance, the 'bolting-on' of sustainable engineering features to existing environmentally flawed systems.

The redesigning and remaking must be carried out at all scales, from the locality in which the action takes place or the built system is sited (such as climate, soils, and plant and animal systems) to connect and protect irreplaceable landscapes from the scale of a site, to a city, to a biome, to the scale of global natural cycles.

Ecological reinventing

One of the key tasks in Ecological Design is 'ecological reinventing', defined here as the process of creating of new and reconstituted human-made systems that function benignly in Nature. Ecological reinventing requires the total rethinking and redefining of what an ecologically effective and sustainable built environment should be and what attributes it must have to be an ecology-driven creation.

Ecological reinventing means developing an improved, redesigned human-made world that works in tandem and in synergy with Nature's systems and processes, and in its interface with Nature, lies within Nature's limits of tolerance. Rather than humanity seeking to control natural systems and processes, the goal is for humanity to work with Nature in a mutually beneficial symbiotic way. It means changing and managing our physical interventions and physical systems, including our societal, socio-economic and political systems, so that they give due deference and respect to Nature, rather than seeking to change it.

While Ecological Design needs to work in a compatible way with technological and engineered systems, a 'technocentric' approach needs to be tempered and adapted and modified to work with and to work within Nature.

Reinventing ecologically means careful consideration of all stages of manufacture and building so that there is no harm to the receiving environment and, once again, wherever possible, the expansion of positive impact to achieve net ecological gain. This reinventing must also include redressing the adverse effects we have already inflicted on Nature as far as practically possible.

Ecological remaking

Ecological remaking involves not just creating new built systems, but reconstructing, rectifying and regenerating *existing* built systems that are not ecologically effective. Our existing and new built environments need to be remade not just in the physical material sense, but in the processes, techniques and the algorithms involved. Remaking follows from the task of redesigning and reinventing.

Ecological recovering and restoring

Ecological recovering and restoring refers to addressing the impairments already inflicted on Nature, regenerating and rejuvenating the ecosystems and the global cycles and systems (and not causing any further impairments). This part of the sustainability equation is fundamental as it restores essential natural capital, increases ecosystem resilience and capacity to adapt to change, and recovers ecosystem services, being the benefits that humanity derives from the presence and functioning of ecosystems and the resources they provide (see Chapter 3).

Even if such restorative action were to start everywhere today, we would still be running a race against the ongoing damage we are still causing daily and against the consequences of damage we have caused in the past. One good example of this that is often ignored by society is the scourge of physiological damage caused by pollutants we discharge to the environment that become hormonal mimics and cause disruption to faunal endocrine systems, including our own.

The concern is that whatever we do, we are already committed to very serious impact on a large portion of Earth's ecological life support base. Accordingly, we must address not just the consequences of ongoing polluting emissions on ecosystems and global cycles, but also the continued presence of existing contaminating discharges into the aerial, aquatic and terrestrial environments.

The noble endeavour of ecosystem restoration is that of attempting to bring degraded ecosystems back to a healthy and 'fully functional' state closer to their reference condition. Some restoration efforts are wilderness-oriented, such as reintroducing lost species to parts of their former distribution, often by undertaking habitat restoration. But ecosystem restoration ranges from large-scale recreation of entire ecosystems down to small-scale reclamation for Nature of urban patches. Whatever the scale, the goal is to restore the landscapes that once existed, along with as many of their functions as possible.

Further complexity in ecological restoration work results from the fact that the 'goal posts' move, both with the innate dynamism of the natural environment and, especially now, with climate change.[39,40] Choosing the reference point for restoration or creation is therefore not straightforward.[41,42] There often needs to be an element of thinking ahead to a community of plants and animals that will be better able to adapt to the predicted future conditions (to the extent that such predictions can be made).[43]

In practice, ecological restoration is a very difficult and time-consuming undertaking, fraught with risk of failure, or at least only partial success. Moreover simply providing vegetation is not the same as ecological habitat creation or restoration: for example, planting trees in an urban area along a street does not create anything much akin to a natural ecosystem. Natural systems require complexity and diversity to thrive. The nutrients in a natural ecosystem flow through the food chain in a 'trophic structure', a cycle of nutrients (see Chapter 3). It clearly follows that it is also not possible for humanity to re-engineer Nature and its biosphere-wide ecological and biogeochemical systems *in their entirety* (though to listen to some technocrats, one might sometimes think otherwise).

Moreover, redesigning, reinventing and remaking the human-made world is not complete without 'closing the loop' and recovering the outputs of making and operating the human-made world. Accordingly, 'recovering' includes addressing what happens to the designed system, its components and materials at the end of their useful and 'used' life, and considering whether all of its components and materials can become part of a new process of design. Alternatively, at the exhaustion of their useful life, materials need to be reintegrated benignly into the natural environment.

Ecological repairing of the broken ecological nexus and reconnecting built and natural systems

Ecologically reconnecting and biointegrating the built environment with Nature means repairing the 'broken link' between our built environment and Nature. The task is to reconnect, physically, biologically and systemically, the interface between the human-made world (and in particular the urban built environment) and Nature – forming what can be called a new '**nexus**'. This must ensure that those parts of the human-made world's urban systems that have become alienated and separated from Nature or are parasitically dependent upon it become physically, systemically and biologically connected and synergistically integral with Nature. We need to recognise the importance of habitat connectivity across all scales. To sustain biodiversity, ecological function, climate resilience, climate mitigation and other ecosystem services, conservation must transcend boundaries and move beyond a site-specific, parcel-by-parcel approach to the scale at which Nature functions.

A key function of Ecological Design is to reconnect the broken physical and systemic connectivity in Nature caused by our ever-expanding built environment and by human fragmentation of ecosystems. On Earth as it exists today, physical and systemic linkages between Nature and the urban environment are typically poor. Disconnection starts with the physical and spatial displacement of the ecosystem upon which an urban area is located, especially when imposed without consideration of the prior ecological conditions of the development site and its ecologically relevant surroundings.

BEING ECOLOGICALLY EFFECTIVE

How can we regard the success of an Ecological Design? The current ways in which humanity **designs, makes** and **uses** its built systems and artefacts are flawed and need to be changed to become **ecologically effective**. 'Ecologically effective' here means causing no irreversible harm to Nature. More specifically, the critical, but not the sole, metrics of this ecological effectiveness include the extent to which humanity, in the manufacture, use and operation of artefacts and built systems, is able to avert and eliminate negative irreversible impacts on Nature's systems and on the biosphere's cycles, preserve and conserve the Planet's natural resources, secure beneficial and positive impacts upon Nature and address the sustainability equation.

'Ecologically effective' means achieving a resilient, durable and sustainable future for all lifeforms and their environments on the Planet – that is, to effect the successful biointegration of our human-made world, especially our built environment, with Nature. This requires reinventing and remaking our existing human-made world and societies into hybrid living systems that function physically connected and systemically fused with Nature and within Nature without dislocation of natural systems in as close an alignment as is practically achievable, moving towards the ecologically perfect world.

'ECOTOPIA' AS HUMAN SOCIETY'S IDEALISED VISION OF ITS FUTURE

What defines our imagined vision of the combination of an ecologically perfect human-made world and built environment and a perfectly benignly organised human society? This is referred here as our **'Ecotopia'**, the antithesis of which is an ecological dystopia. Ecotopia is a world that has clean air, clean water, clean land, healthy ecosystems, clean built environments that enable healthy ecosystems and a healthy and happy human society to thrive.[44]

Specifically, 'clean air' means an aerial environment that is free from anthropogenic pollution by hydrocarbons, excess oxides of sulphur and nitrogen, ozone, lead and a host of other contaminants.

'Clean water' means an aquatic and marine environment that is free from anthropogenic contamination by chemicals, pathogenic microorganisms and unrecycled disused items such as waste plastics.

'Clean land' means, for example, a terrestrial environment that has healthy productive soils that have not been degraded by excessive working for agriculture or applications of chemical fertilisers, herbicides and pesticides, which is in ecological balance so that parasites and pathogens are not rife.

A 'clean built environment' is one whose systems and processes do not contaminate or do irreversible harm to Nature and to human society, whose production and operational systems and processes are not deleterious to natural ecosystems and global cycles, and which functions on renewable systems of energy.

'Healthy ecosystems' are ecosystems where their ecological integrity, carrying capacities and assimilative capabilities remain intact and are infinitely capable of providing ecosystem services.

An ecologically 'healthy human society' is one in which its inhabitants' well-being, ways of life and social-economic-political-institutional systems are non-disruptive of Nature and exist in a harmonious integral partnership with it – with opportunities for physical activity and mental respite – all of which factors foster human health and well-being.

In Ecotopia, the 'logistics' of the materials used in the built environment are cyclical, and not 'throughput'. At the eventual exhaustion of each material's used and useful life, all those remaining materials that are no longer reusable or recyclable or which cannot be remade or reconstituted and transformed into forms for a new process of design and use will then need to be benignly returned back to Nature, such as by biodegradation or by biointegration with potential for their replenishment.

In Ecotopia, ecologically inspired living systems abound, strengthening human links with Nature immeasurably, fostering stewardship, understanding and symbiosis, and creating a future both for humanity and all other forms of life on Earth. Ecotopia reverses many of the values with which humanity has been indoctrinated, which have tended to physically and systemically distance humankind from Nature to foster human interventions that unsustainably exploit Nature and its resources.

Moreover, in Ecotopia, humanity stops designing for the continued production of goods and artefacts that are ecologically detrimental to the environment, and which are not vital for human existence and an enduring civilisation. The fault lies in existing commercial business enterprises and marketing endeavours that continue to encourage human society to acquire more and more artefacts beyond those essential for its existence, and which suggest that Planet Earth is an infinite source of material and energy resources for humanity's use and consumption. The need for working within the Planet's limitations or natural resources and ecosystem services extends to the Planet's limited ability to function as an environmental sink for humanity's wastes, and within its other limiting resilient aspects.

In this vision of Ecotopia, the built environment and all of its technological systems do not exist and function as entities separate from the natural environment, so that there are 'city–country' or 'urban–exurban' dichotomies, but rather as integrated techno-biological systems. Ecotopia is a world where Nature, human society and technology become inextricably and harmoniously linked. In Ecotopia, human society no longer regards Nature and its built environment and machines as separate entities, as 'us versus Nature', but the functions in Nature and those of the built environment are merged so that the boundaries between Nature and the built environment are completely blurred and fused and not in conflict, whereby their functioning becomes systemically indistinguishable from one another. Technology, engineering and machines become integrated as part of the biotic and abiotic constituents of Nature. Our thinking becomes a hybrid of biological and non-biological thinking.

In Ecotopia, we also address past impairments of the natural environment, repairing and rejuvenating damaged ecosystems and biogeochemical cycles, repairing the dislocations in the intersect between the physical and systemic aspects of the existing built environment with the flow and processes of the Nature's systems, and enabling Nature and the human-made world to coexist and thrive together in a harmonious, mutually beneficial and compatible partnership on a healthy Planet over time.

Ecotopia is, then, an ideal and a set of aspirations to be achieved by human society. To date, there is no existing human-made system and built environment that can approximate to the condition that can regarded as our Ecotopia. Many of the current efforts at Ecological Design are only partially ecologically effective and successful. For instance, many do not biointegrate effectively nor synergistically in totality and benignly with Nature's systems.

Furthermore, we are currently still not very good at designing for evolving and dynamic environments. In Ecotopia, our relationship with

Nature is not to be a static state and juxtaposition, but rather a dynamic partnership, responding to local and global environmental and sociocultural change.

In essence, achieving Ecotopia is humanity's ultimate objective as a vision that we seek to achieve. This, then, is the challenge of accomplishing the tasks set out above in the endeavour to 'save the Planet'. The reinventing and remaking of the existing built environment need to be in as close an alignment as possible with each other and with the ideal of Ecotopia. In the practice of business in the commercial world, this is referred as a Strategic Plan. Achieving Ecotopia is our Mission Statement.

ADDRESSING THE ENVIRONMENTAL PROBLEMS WE HAVE CREATED?

What can we do to address the environmental problems we have created? It is argued here that we must start by reconsidering the ways we contrive and construct the existing built environment. Achieving this biointegration must start with changing from our approach to Nature.

We need to avert the current predominant approach and instead seek a seamless, synergistic, symbiotic, mutually beneficial, co-existent relationship with Nature's ecosystems and its biogeochemical cycles where all of humanity's acts seek to conserve, protect and benefit natural ecosystems and habitats in the presence of human activities and development.

Whereas in the immediate past our actions needed to be preventive, the situation today is one of a 'race and rescue' mission. Ecological Design is relevant because the endeavour to save the Planet is the most compelling issue to be addressed today, not only by all designers and all those whose daily work impinges on the environment, but by all of humanity, as all are affected.

We must change from our current approach of predominance over Nature and instead seek a seamless, synergistic, symbiotic, mutually beneficial, co-existent relationship of our built environment with Nature that is referred here as 'effective biointegration'. In this endeavour, we need to protect and benefit natural ecosystems and habitats in the presence of human activities and development.

If we are able to achieve these goals successfully, we may perhaps be able to curtail our ongoing depredations of the global environment and begin to repair the damage we have already caused. This is the challenge humanity must successfully meet for its sustainable future and the sustainable future of the many other wonderful and unique species for which we have now inherited a moral duty of care and stewardship. This is the goal and challenge for Ecological Design.

Accomplishing the tasks stated here will not be a simple, forthright endeavour – the premise is that Ecotopia is the green and sustainable world and built environment combined with an equable and sustainable global society that most of us want, the world we must make – and a world that is achievable. This declaration of intent (as a 'manifesto') is the ambitious and audacious goal of Ecological Design.

NOTES

1 Berner, R. A. *The Phanerozoic Carbon Cycle: CO_2 & O_2*. Oxford University Press, 2004.
2 Environmental Literacy Council. 'Biogeochemical Cycles.' 2018, https://enviroliteracy.org/air-climate-weather/biogeochemical-cycles/.
3 Cain, M. L., et al. *Ecology*. Sinauer Associates, Inc. Publishers, 2014 (Ch. 22).
4 Bunker, A. F. 'Computations of Surface Energy Flux and Annual Air–Sea Interaction Cycles of the North Atlantic Ocean.' *Monthly Weather Review*, vol. 104, no. 9, 1976, pp. 1122–1140. doi:10.1175/1520–0493(1976)1042.0.co;2.
5 Environmental Literacy Council. 'Sources & Sinks.' 2018, https://enviroliteracy.org/air-climate-weather/climate/sources-sinks/.
6 Karl, T. R. *Global Climate Change Impacts in the United States*. Cambridge University Press, 2009, pp. 13–52.
7 Mathez, E. A., and Smerdon, J. E. *Climate Change: The Science of Global Warming and Our Energy Future*, second edition. Columbia University Press, 2018, https://blogs.ei.columbia.edu/2018/11/14/a-new-primer-on-climate-change/.
8 Intergovernmental Panel on Climate Change. *Special Report on Global Warming of 1.5 ° C (SR15)*. October 2018, www.ipcc.ch/.
9 Environmental Literacy Council. 'Carbon Cycle.' 2018, https://enviroliteracy.org/air-climate-weather/biogeochemical-cycles/carbon-cycle/.
10 Reisner, M. *Cadillac Desert: The American West and Its Disappearing Water*. Penguin, 1993, pp. 1–47.
11 Singh, V. P., Mishra, A. K., Chowdhary, H., and Khedun, C. P. 'Climate Change and Its Impact on Water Resources.' In L. K. Wang and C. T. Yang (eds), *Modern Water Resources Engineering*. Humana Press, 2014, pp. 525–569.

12 Costanza, R., et al. *Ecosystem Health: New Goals for Environmental Management.* Island Press, 1992.
13 Lu, Y., et al. 'Ecosystem Health towards Sustainability.' *Ecosystem Health and Sustainability*, vol. 1, no. 1, 2015, p. 2. doi:10.1890/EHS14-0013.1.
14 Rapport, D. 'Assessing Ecosystem Health.' *Trends in Ecology & Evolution*, vol. 13, no. 10, 10 Oct. 1998, pp. 397–402., doi:10.1016/s0169-5347(98)01449–9.
15 Gunderson, L. H. 'Ecological Resilience: In Theory and Application.' *Annual Review of Ecology and Systematics*, vol. 31, no. 1, 2000, pp. 425–439, www.annualreviews.org/doi/abs/10.1146/annurev.ecolsys.31.1.425?journalCode=ecolsys.1.
16 Ocean Health Index, 'Ecological Integrity.' 2019, www.oceanhealthindex.org/methodology/components/ecological-integrity.
17 Darwin, C. *On the Origin of Species*. John Murray, 1859.
18 Willis, K. J., and McElwain, J. C. *The Evolution of Plants*, second edition. Oxford University Press, 2014.
19 Cain, Michael L., et al. *Ecology*. Sinauer Associates, Inc. Publishers, 2014 (Ch. 17).
20 Harris, J. 'Soil Microbial Communities and Restoration Ecology: Facilitators or Followers?' *Science*, vol. 325, p. 573 (July 2009).
21 Didham, R. K. 'Ecological Consequences of Habitat Fragmentation.' In *Encyclopedia of Life Sciences*. John Wiley & Sons, 2010, www.els.net.
22 Lal, R. 'Carbon Sequestration.' *Philosophical Transactions of the Royal Society B: Biological Sciences*, vol. 363, no. 1492, 30 Aug. 2007, pp. 815–830. doi:10.1098/rstb.2007.2185.
23 Finlay, J. C., et al. 'Human Influences on Nitrogen Removal in Lakes.' *Science*, vol. 342, no. 6155, 11 Oct. 2013, pp. 247–250. doi:10.1126/science.1,242,575.
24 Wootton, J. T., et al. 'Effects of Disturbance on River Food Webs.' *Science*, vol. 273, no. 5281, 13 Sept. 1996, pp. 1558–1561. doi:10.1126/science.273.5281.1558.
25 Liss, K. N, et al. 'Variability in Ecosystem Service Measurement: A Pollination Service Case Study.' *Frontiers in Ecology and the Environment*, vol. 11, no. 8, 26 June 2013, pp. 414–422. doi:10.1890/120189.
26 DeLucia, E. H., et al. 'Net Primary Production of a Forest Ecosystem with Experimental CO_2 Enrichment.' *Science*, vol. 284, no. 5417, 1999, pp. 1177–1179. JSTOR: www.jstor.org/stable/2899039.
27 Pugh, T. A. M., et al. 'Role of Forest Regrowth in Global Carbon Sink Dynamics.' *Proceedings of the National Academy of Sciences*, 19 Feb. 2019, doi:10.1073/pnas.1,810,512,116.
28 Peterson, G. D., Allen, C. R., and Holling, C. S. 'Ecological Resilience, Biodiversity, and Scale.' Nebraska Cooperative Fish & Wildlife Research Unit – Staff Publications, vol. 4, 1998, http://digitalcommons.unl.edu/ncfwrustaff/4.
29 Stockholm Resilience Centre, 'The Nine Planetary Boundaries.' 2015, www.stockholmresilience.org/research/Planetary-boundaries/Planetary-boundaries/about-the-research/the-nine-Planetary-boundaries.html.
30 Despommier, D. D. *The Vertical Farm: Feeding Ourselves and the World in the 21st Century*. Thomas Dunne Books, 2010.
31 Gough, C. M. 'Terrestrial Primary Production: Fuel for Life.' *Nature Education Knowledge*, vol. 3, no. 10, ser. 28, 2011, p. 28, www.nature.com/scitable/knowledge/library/terrestrial-primary-production-fuel-for-life-17567411.
32 World Bank, 'Part III: Cities' Contribution to Climate Change.' *Cities and Climate Change: An Urgent Agenda*. World Bank, 2007, pp. 14–32, http://siteresources.worldbank.org/INTUWM/Resources/340232-1205330656272/4768406-1291309208465/PartIII.pdf.

33 Kaminski, L. A., et al. 'Interaction between Mutualisms: Ant-Tended Butterflies Exploit Enemy-Free Space Provided by Ant-Treehopper Associations.' *The American Naturalist*, vol. 176, no. 3, 20 July 2010, pp. 322–334. doi:10.1086/655427.
34 Cain, M. L., et al. *Ecology*. Sinauer Associates, Inc. Publishers, 2014 (Ch. 14–15).
35 Wirsen, Carl O., et al. 'Chemosynthetic Microbial Activity at Mid-Atlantic Ridge Hydrothermal Vent Sites.' *Journal of Geophysical Research*, vol. 98, no. B6, 10 June 1993, p. 9693. doi:10.1029/92jb01556.
36 Karamouz, M. *Hydrology and Hydroclimatology: Principles and Applications*. CRC Press, 2013.
37 Margulis, S. *Introduction to Hydrology*, 4th ed. 2017, https://margulis-group.github.io/teaching/.
38 Dang, T. D., et al. 'Hydrological Alterations from Water Infrastructure Development in the Mekong Floodplains.' *Hydrological Processes*, vol. 30, no. 21, 29 Apr. 2016, pp. 3824–3838. doi:10.1002/hyp.10894.
39 Bao, K., and Drew, J. Traditional Ecological Knowledge, Shifting Baselines, and Conservation of Fijian Molluscs. *Pacific Conservation Biology*, vol. 23, 2017, pp. 81–87, https://doi.org/10.1071/PC16016.
40 Duarte, C. M., et al. Return to *Neverland*: Shifting Baselines Affect Eutrophication Restoration Targets, *Estuaries and Coasts*, vol. 32, 2009, p. 29, https://doi.org/10.1007/s12237-008-9111-2.
41 Heymans, J. J., et al. 'Global Patterns in Ecological Indicators in Marine Food Webs: A Modelling Approach.' *PLOS ONE*, 2014, https://doi.org/10.1371/journal.pone.0095845.
42 Olson, R. 'Slow Motion Disaster below the Waves.' *LA Times*, Sunday Opinion Section, Environment. 17 Nov. 2002, www.shiftingbaselines.org/op_ed/.
43 Papworth, S. K. et al. 'Evidence for Shifting Baseline Syndrome in Conservation.' *Conservation Letters*, vol. 2, 2009, pp. 93–100, https://onlinelibrary.wiley.com/doi/pdf/10.1111/j.1755-263X.2009.00049.x.
44 More, T. *Utopia* [1516], translated by K. P. Marshall. Washington Square, 1965.

FURTHER READING

Birkeland, J. 2002. *Design for Sustainability: A Sourcebook of Integrated Ecological Solutions*. Earthscan, London.

Desai, P. 2009. *One Planet Communities: A Real-Life Guide to Sustainable Living*. John Wiley & Sons, London.

Edwards, B. 2005. *Rough Guide to Sustainability*. 2nd edition. RIBA Publishing in Association with Earthscan, London.

Farr, D. 2008. *Sustainable Urbanism: Urban Design with Nature*. John Wiley & Sons, Hoboken, NJ.

Giradet, H. 1999. *Creating Sustainable Cities (Schumacher Briefing)*. Green Books, London.

Low, N., Gleeson, B., Green, R. & Radović, D. 2005. *The Green City: Sustainable Homes, Sustainable Suburbs*. Routledge, Abingdon, UK

Register, R. 2006. *Ecocities: Rebuilding Cities in Balance with Nature*. New Society Publishers, Gabriola Island, Canada.

Rogers, R. 1997. *Cities for a Small Planet*. Faber & Faber, London.

Ruano, M., ed. 1998. *EcoUrbanismo*. Gustavo Gili, Barcelona, Spain.

William McDonough & Partners. 1992. *The Hannover Principles: Design for Sustainability*. William McDonough & Partners, Charlottesville, VA.

TUCK

2

Redefining design to include the ecological sciences

The principle of ecocentricity

DEFINING ECOLOGICAL DESIGN: THE PRINCIPLE OF ECOCENTRICITY

What differentiates Ecological Design from traditional design? Designing to address environmental issues of requires a different approach than conventional practice. We can define Ecological Design as the synthesis and resolution of the ecological determinants and factors in a design problem while endeavouring to achieve a resilient, durable and sustainable future for all life forms and their environments on Earth. The difference is that the conceptualisation of the designed system, the resolution of all the usual factors in design in the problem-solving endeavour of designing and the subsequent production of the designed system itself are influenced and guided at every stage by the **science of ecology** and **environmental biology** (see Glossary) as the branch of biology that deals with the relations of organisms to one another and to their physical surroundings. Adopting the science of ecology in designing is what is missing in most approaches to sustainable design by other green designers. Adopting the science of ecology as the guiding and determining factor might even be further stated as an obvious factor as, after all, if we are to protect and restore and conserve the health of the Planet, in effect this involves protecting, restoring and conserving the Planet's ecology. This principle of adopting the science of ecology is referred to here as '**ecocentricity**'. This is one of the significant and important principles presented here in our drive to save the Planet.

The ecocentric design approach involves seeking to minimise, or ideally eliminate, any potentially negative consequences of humanity's actions and its designed systems on all of the Planetary 'spheres', namely the **atmosphere, biosphere, hydrosphere** and **geosphere** (see Glossary), and furthermore seeks beneficial and positive outcomes for Nature at all stages in the design process.

The approach requires and can start with the advance assessment of the likely impacts of any designed system or human action on natural and **semi-natural ecosystems** (see Glossary) and biogeochemical cycles. This assessment includes considering all organisms and their interactions within their environment, including the impact of a human activity or of the designed system or artefact, both on site and distantly removed.

This consideration needs to be extended to the environmental consequences of all uses of materials in the production of the designed system in their logistics and flow from each of their sources to their sink or eventual biointegration back into the environment. Clearly, in Ecological Design the consideration includes many more factors as being pertinent to the *design problem*, to the designed system and its content than in the traditional non-ecological approaches, in which all must work with an in-depth knowledge of ecology and an understanding of the designed system's life-cycle.[1]

In undertaking Ecological Design, the starting ideology must be such that we do not perceive Nature as an infinite and limitless source of natural resources, of ecosystems and ecosystem services, to be used and consumed at will. Instead, it seeks to prevent and to avoid excessive consumption and depletion of the Planet's finite non-renewable energy and material resources, and all other consequences of detriment to Nature and the Earth. Ecological designers must in all instances seek to mitigate and eliminate negative impacts on Nature and the degradation of Nature's ability to provide ecosystem services.

Ecological Design is also not a solely anthropocentric approach that is merely aimed at ensuring humanity's long-term survival and the flourishing of human civilisation. There is also – and this is a fundamental issue overlooked or ignored by so many designers and interventionists today – a vital *moral and ethical* obligation towards life on Earth that comes with our special ability to analyse, predict and affect our Planet to the extent that we do. Adopting an ecocentric approach is regarded as essentially 'the right thing to do'. With each project that follows this approach, we move closer to being a steward of the Planet's biodiversity, systems and natural resources and away from humanity being a presumptuous and voracious exploiter and despoiler of the web of life of which we form but part.

It follows that in implementing Ecological Design, it is necessary to embrace the considerable challenge of working with the great complexities both of human society with all its social-economic-political-institutional and technological systems, and with Nature in all its even greater complexity.

DEFINING DESIGN AND ITS PROCESS

What is the process of '**Ecological Design**'? As emphasised earlier, the focus of the redesigning and remaking needs to be based on the objective of achieving effective biointegration of our human-made world and its built environment with Nature. However, within the context of Ecological Design, we need first to define what we mean by '**design**'.

'Design' is regarded here as the practical problem-solving approach for the synthesis and resolution of a set of factors and determinants involved in creating something new. It is the process of seeking solutions based on abductive reasoning, culminating in a solution that can take the shape of a physical form, a concept, a system, a process or even a way of thinking. This designed output may be represented, for example, in a visual medium such as a drawing that provides a representation and model of reality. In a rigorous approach, the representation is then realised in the form of a physical or virtual prototype, which needs testing and refining before the final design is produced and adopted for use by human society to meet the social, political and aspirational requirements the design must satisfy. The process of designing and conceptualising a solution is a 'problem-solving' process. Design approaches relating to objects and places have been adopted for use by other disciplines, such as business management.

THE TECHNOCENTRIC APPROACH AND ITS SHORTCOMINGS

What are the other current approaches to sustainable design? While many would see merit in the arguments for an ecocentric approach (if not always being prepared to apply it with rigour), the current human generation tends to behave as if highly devoted to human technological advances and achievements while ignoring Nature.

Technology clearly has huge benefits (and can be made to work synergistically with Nature, as will be discussed later). However, society's existing social-economic-political-institutional systems have encouraged the autonomous, unchecked development of its technologies often without considering the ecological consequences.

Moreover, the present generation of sustainable design and much of humanity tend to place immense faith in humanity's technology and engineering systems and humanity's innovative prowess to address the sustainability equation (see Chapter 1). There are now those who argue that Nature's intrinsic resilience has been so extensively compromised and reduced that humanity must turn to its technological systems to re-engineer Nature and to reconfigure the Planet's metabolism. A fully technocentric approach, however, would be premised on the erroneous belief that Nature's functionalities can be replaced fully by technology and engineering. Such an approach, however, is fundamentally flawed and highly anthropocentric.

Compared to the complexity and intricacy of natural ecosystems and living systems, even the most complex abiotic human-made systems are relatively simple. All life on Earth is *interdependent* and is deeply bound in the dynamic, complex and indeterminate natural network of relationships. Natural ecosystems have interactive complex pathways that are highly co-dependent. Any change and manipulation of any one fragment or part of an ecosystem causes a reverberation throughout the system. Elements within natural ecosystems then act to bring them back to equilibrium again, and can do this as long as environmental perturbation is not too great (this is the 'homeostatic' property of ecosystems).[2]

All of humanity's actions and interventions have effects on the Planet's natural systems, whether these are negative ('harmful'), neutral or positive. For instance, introducing a plant species which is alien to a particular locality can in some cases lead to the rapid spread of this species, hence its introduction could potentially be to the detriment of local habitats, flora and fauna.[3,4]

A seemingly innocuous act such as a human switching on the lights at home contributes to processes that are driving global climate change if the energy source is fossil fuel-based. Similarly, every time a person is driving a car, that person is contributing to changing the climate, acidifying the oceans and melting the ice caps, with impacts on the natural environment and Nature's systems that support life on the Earth.[5]

Consider a technological and engineered system such as a train engine. The engine can be taken apart and put back together again to resemble closely the original, whereas ecosystems cannot be taken apart and be put together again. The train can be stopped and moved backwards and restarted, whereas ecosystems, while often able to recover from environmental shocks, cannot be simply stopped, reversed and restarted.

A technological and engineered system can have any of its defective parts replaced to revert to its original state of functioning without any difference from its original condition, whereas an ecosystem cannot be readily taken apart and reassembled. A human-made technological and engineered system can be assembled out of pre-existing parts, whereas an ecosystem as seen in Nature cannot be assembled in full out of pre-existing parts.

Habitat creation approaches can commence an ecosystem from scratch by combining a variety of biotic and abiotic components, but the system then grows and develops further modified by natural colonisation, co-developed evolutions and more such that the resultant system will always have many more unexpected 'emergent' properties than any technological system.

Acknowledging these aspects in technological systems, we can and must still address the creation of hybrid systems that are techno-biological systems designed to be as seamless a collaboration and augmentation of technology with biological systems as possible (see Chapter 3).

Conventional certification and accreditation systems for the built environment, such as LEED, BREEAM and Green Mark, have had some notable impact in informing designers and the general public about the principles, ethics and practice of sustainable design. However, many designers who have gained in-depth knowledge from working with these 'green rating' systems as are currently practised, after having become well-informed and advanced in their use and having delivered several highly accredited 'green-rated' buildings and masterplans (such as with 'platinum' rating), then question whether just designing to meet the criteria of accreditation systems is 'all there is' to green design.

Most of the existing accreditation systems tend to be prescriptive, not always performance-based. More importantly, they do not comprehensively consider the biosphere and ecology in their criteria. This is what differentiates Ecological Design from simply design-based compliance with an accreditation system. Existing accreditation systems need to be extended to encompass a comprehensive ecology-centric approach to include full ecological considerations in designing and making the built environment and technological artefacts. They should be further extended to include human society's social-economic-political systems as one of the crucial 'infrastructures' (see Chapter 8) to be integrated into an overall holistic urban design.

THE ANTHROPOCENTRIC APPROACH: ADDRESSING SOCIETAL NEEDS AND VALUES, BIOPHILIA AND HUMAN WELL-BEING

Should Ecological Design be anthropocentrically focused? The answer is 'yes', while still remaining cognisant of and being responsive to the natural environment and its constraints. Anthropocentricism alone is another similarly flawed approach to design, as it focuses entirely on fulfilling the needs of human society and ignores the needs of wider Nature. We can contend that it is the over-emphasis on the needs and requirements of human society while ignoring Nature that is the root cause of much of existing environmental impacts.

In Ecological Design, we need to recognise that human beings are but one of the many species that have evolved alongside each other. Each species, however seemingly insignificant, has a current or latent potential role in the functioning of the biosphere. If we continue along the path of fulfilling selfishly only our human society's needs, mindless of natural consequences, we will not only achieve a great destruction of the 'library of life', but will almost certainly deprive future generations of opportunities to meet their needs, condemning them to an impoverished, collapsing biosphere.

ECOLOGICAL DESIGN AS APPLIED ECOLOGY

How is Ecological Design similar to practices in environmental conservation? Ecological Design is in effect a logical extension of 'applied ecology'. The traditional methods of applied ecology – the practical applications of ecology to the challenge the management and conservation of the Planet's biological and physical resources – are now being extended to the designing of humanity's built environment.

Ecological Design seen in this light is the practical application of ecology to manage environmental consequences at various scales of interaction, from the macro level, such as impacts on global processes, to the micro level, such as the local spatial displacement of flora and fauna by a built structure. This further applies to the logistics of materials used in the designed system over the life-cycle of each material.

ADDRESSING THE ISSUE OF AESTHETICS IN ECOLOGICAL DESIGN

How can we reconcile the need to address the aesthetic requirements in designing with the science of ecology? Aesthetics concerns the way something looks. Fulfilling aesthetic design outcomes can mean creating designs that are multi-sensorial experiences for human users. How a designed system looks is the first thing that affects human society's reaction to a designed system, and influences whether it is accepted and widely adopted by human society or rejected.

Design, besides being a problem-solving process that is iterative and exploratory, is also an artistic, creative and partly intuitive endeavour that is as much an art as it is a technical process, whereas ecology is a science. Although the aesthetic qualities of a designed system are subjective and indeterminate, they play an important and vital role in Ecological Design.

Ecological Design must then technically challenge the inclusion of 'unnecessary' or gratuitous forms and articulation. We need to avert the situation where the designer designs with his or her own personal and frequently merely gratuitous discourse in form-making and intellectual design emphases that are not ecologically benign. The aesthetic articulation and form-making in Ecological Design need to be justified by environmental benefits to Nature. Ecological Design needs to avoid the addition of physical features and technological systems to a designed system in a way that is unnecessary and results in a net waste of the Earth's finite material and non-renewable resources. The question that must be constantly asked when designing is whether the design configuration and aesthetic articulation are unnecessary and will add unacceptable wastes in the designed system that might impair the integrity of ecosystems or the functioning of any global processes. The aesthetic articulation in Ecological Design must be justified by ecological benefits, such as in effecting energy savings in the use of non-renewable sources of energy. If the balance of effects is net negative, then perhaps the designed system's articulation should be not adopted, or the designed system itself should redesigned or not be made at all. In addressing the aesthetics of any designed system, Ecological Design needs to achieve a sensitive balance between the core functionality and primary purpose of the design, ecological considerations (avoiding net harm and seeking net environmental gain) and securing biophilic benefits. In effecting these, Ecological Design must ensure that the various functionalities are appropriately balanced and that any apparent adoption of one articulation or feature does not unduly detract from optimising delivery of another.

Ecological Design must challenge the marketing approaches of unethical producers in human society who capitalise on human addictions to novelty and competition that lead to gratuitous waste. A good example of this is the trend for the latest fashionable electronic gadgetry which contributes to the growth in annual discarding of staggering volumes of electronic waste (e-waste). The average American consumes 4100 lbs of e-waste over his or her lifetime.[6] Over 40 million tonnes of primarily completely functional e-waste were generated in 2016.[7] This is 8% more than in 2014 and twice the rate of increase of plastic waste. This (largely unnecessary) mountain of refuse would tip the scales against 4500 Eiffel Towers. This does not mean that a design should be bland and without delight, but it means that design articulation should be justified by environmental or ecological rationales.

A possible approach in addressing of the aesthetics of Ecological Design might be through '**biophilia**',[8] where the biotic aesthetics create multi-sensorial experiences for human users and enhance human users' well-being. In recent years, the theory of biophilia has emerged and has been substantiated by an ever-increasing body of empirical evidence. In brief, biophilia theory maintains that humankind has an innate affinity for Nature – an affinity that applies across different races, ages and gender. Cited manifestations of this include our desire for green cities, vegetated buildings, to create diverse gardens and to include natural design patterns. Developed from this theory is the emergent practice of **biophilic design**. This entails detailed consideration throughout the design process of those aspects of Nature with which we have the greatest affinity in attempt to maximise their manifestation and positive impacts on human well-being. Evidence shows that when our positive affinities for Nature are fed and satisfied and our associated biophobic tendencies allayed (generally through education), we can benefit in terms of psychological health and well-being in many different ways, and very significantly.

For example:

- Living in or proximity to open green spaces and protected natural areas can enhance human functioning and well-being, notably when compared to living in urban environments and in hardscapes with little or no access to such naturalistic settings. This appears to be particularly true when such areas are biodiverse, characterised by the manifest presence of clean, fresh and preferably flowing water, provide refuge and vantage to the users of the space and which are often savannah-like (semi-wooded) in terms of vegetation cover.

- Neighbourhoods and communities characterised by superior natural environmental quality tend to foster more positive environmental values, reduced crime rates and higher quality of life.
- Neighbourhoods lacking such features appear more likely to foster crime, vandalism and social pathologies.
- The experience of Nature can actually notably accelerate human healing and recovery from illness and surgery.
- Contact with Nature can foster social ties and relationships.
- Contact with Nature can contribute to enhancing intellectual performance and problem-solving.
- Work settings that are biophilically designed can reduce stress, enhance performance and increase productivity.

It is necessary to recognise that while such benefits appear universal, they can be of particular value to those in society who are disabled, ill, isolated, disadvantaged or generally vulnerable.

In addressing the aesthetics of any designed system, Ecological Design needs to achieve a sensitive balance between the core functionality and primary purpose of the design, ecological considerations (avoiding net harm and seeking net environmental gain) and securing environmental as well as biophilic benefits. In doing this, the designer must ensure that functionalities are appropriately balanced and that obsession with one does not unduly detract from optimising delivery of another.

LOCAL INFORMANTS OF ECOLOGICAL DESIGN

What are the key informants of Ecological Design? A process of iterative evaluation of ecological impacts through the design process and the subsequent synthesis is absolutely key to the process of Ecological Design. Environmental consequences should be assessed at each step in the flow of the materials and energy resources to be used in the making, use, recovery and reintegration into Nature of any designed system.

This process is essential if we are to minimise the risk of unforeseen detrimental environmental outcomes. In fact, ecological impact assessment should be the key determinant not just for designing, but for all of humanity's actions and interventions on Earth.

Accordingly, Ecological Design requires the study of the biota, the physical environment and the complex interactive pathways in ecosystems – studies that can become particularly involved in relation to pristine and sensitive ecosystems. Particular topics of assessment for the ecological designer, deriving from the primary studies of ecosystem status and condition at the site of any human intervention, include renewability of resources and the ability of ecosystems to provide ecosystem goods and services. Studies relevant to this can include those investigating:

- Temporal and spatial patterns of biodiversity.
- The flow of energy through ecosystems, the physical circulation of matter through food chains and the cycling of all nutrients aside from food.
- Ecosystem development and evolution and variations in ecosystem productivity.
- The sequestering and assimilation of contaminants and the function of the environment as a sink for anthropogenic waste and discharges.
- Diurnal year-to-year variations in ecosystem processes – for example, biota can accumulate during resource-rich periods, but crash in periods of scarcity, and all associated functionalities will change accordingly.
- Climatic and environmental variations on system performance – for example, a drought, an especially cold winter or an outbreak of pest species can all strongly affect the performance of a designed system that has biotic components, and these effects need to be anticipated as far as possible.
- Impacts at source sites for materials and energy used at a project or product site – for example, extraction of resources

to make and generate the urban areas in which we live often results in significant ecological impacts far from the site of manufacture or use of the designed element of system; these impacts often extend beyond the depletion of non-renewable and finite energy resources and other natural resources, extending to loss of valuable habitats, precious species, and the ability of ecosystems to provide services and goods to humankind in these source localities. A good example of this is marine sand extraction for use in manufacture of concrete to fuel the construction industry.

- Options for control and management of ecosystems, including how such management may become a seamless combination of the biospheric ecosystems and human-made technologies.

ECOLOGICAL DESIGN'S ROLE IN SECURING NET POSITIVE ENVIRONMENTAL GAIN

Can Ecological Design go beyond seeking to mitigate and eliminate negative consequences and also positively contribute to the environment? Ecological Design should extend to securing net positive ecological environmental outcomes and benefits to Nature, to repairing and rejuvenating the natural environment damaged by human intervention and human-made systems. Such an approach extends to sequestering carbon, enhancing appropriate local biodiversity, normalising ecosystem productivity and restoring water quality or soil health, for example through bioremediation, increasing provision of ecosystem services.[9]

The key rationale for this approach is the existing degraded state of much of existing global ecosystems. Our degradation of ecosystems has resulted in a reduction of their resilience to further environmental change. Of greatest concern to environmentalists today is the potentially devastating effect of climate change. By restoring ecosystems, we can also restore long-term resilience of all their components – known as ecosystem 'integrity'. Similarly, if we work to restore degraded ecosystems, we help to replenish the store of finite resources and their ability to keep providing ecosystem services.

It also follows that in Ecological Design we need to ensure that net environmental gain achieved at one stage of the life-cycle of a product or designed area is not undone by adverse effects at another stage.

IMPLEMENTING ECOLOGICAL DESIGN IN AN INFRASTRUCTURE-LED APPROACH

The effective Ecological Design of our built environment cannot be limited to the incremental scale of single buildings or even clusters of buildings, but should be addressed at the larger scale of urban 'infrastructures' (see Chapter 8). By this is meant the systems designed to support and permit the flows of energy, water, biodiversity, people and indeed ideas through the urban fabric. These 'infrastructures' must be interlinked and 'ecologically effective'. In other words, the integration of infrastructures must be such as to give precedence to Nature and the provision of ecosystem services (see Chapter 3).

ECOLOGICAL DESIGN AND MOBILITY

A particular focus of the Ecological Design approach as part of its technological infrastructure is human society's requirement for **mobility** (of people, materials, food and goods, whether by land, sea or air) and the consequences of this in terms of the spatial impacts, consumption and depletion of non-renewable resources and pollution emissions.

In the great majority of current societal and urban conditions, mobility systems remain dependent on motorised vehicles, public transportation, mass-transit rail systems, air travel and other movement systems powered primarily by fossil fuels.[10,11] Accordingly, these systems emit vast quantities of contaminating greenhouse gases that are fuelling current global climate change.[12] Ecological designers today should see humanity's transportation systems as being in urgent need of reinvention.

ECOLOGICAL DESIGN AND EMBODIED IMPACTS

What are the embodied consequences of a designed system? Ecological Design needs to consider the environmental impacts inherent and embodied in every component and material used to make, operate and recycle its built environment, including their embodied energy, carbon emissions and the embodied impacts on the ecosystems and biogeochemical cycles in its flow over its life-cycle.

One general index in Ecological Design to quantify energy implications of materials used in any designed system is its '**embodied energy**' (see Glossary). Embodied energy is the sum of all the energy required to produce any goods or services, considered as if that energy was incorporated or 'embodied' in the product itself – the total of the energy necessary for a designed object or system's entire life-cycle through raw material extraction, transport, manufacture, assembly, installation, disassembly, deconstruction and/or decomposition. This embodied energy can be partly discounted by the carbon benefits of reuse through recovery and recycling.

While this consideration is vital in determining the full cost-benefit analysis of any given designed product or system in terms of potential climate change, it is important that the ecological designer does not become 'a slave to carbon accounting' as the sole determinant of effective Ecological Design. Net carbon emissions must always be minimised (or reversed) as far as possible, but not at the expense of losing other key ecological and socio-ecological systemic functionalities.

For the ecological designer, assessment and evaluation does not end with the completion of designing and production or construction of the designed system. The recovery of materials at the end of their useful and 'used' life has to be considered. This is known as 'closing the loop' in the cycle of materials. 'Used life' refers to materials that cannot be further physically reused in their discarded form, but rather need to be reduced to raw ingredients and then reconstituted as new useful materials, such as with used metal, certain plastics and paper materials.

Ecological Design always looks to options for the *biointegration* of waste materials back into the natural environment, and wherever possible, to enable their eventual replenishment through natural systems.

Ecological Design at the outset must facilitate the future deconstruction, disassembly and reuse of every part of the designed system. This is another aspect of Ecological Design that differentiates it from traditional design. In typically practised design, uninformed by ecology, once the designed system has been produced or constructed, there is often little or no interest with what happens to the designed system at the end of its used life.

Ecological Design should seek to go beyond designing based on **singularly focused** priorities such as 'net zero energy buildings', positive energy buildings', waste recycling systems (domestic and industrial) and other systems and labels.

These are all important single subjects, but these approaches need to be framed within the *much larger picture of holistic sustainable Ecological Design*. Although all of these contribute in part to the overall sustainability equation, they remain ineffective if singularly, disparately and incrementally adopted in addressing aspects of the broader issues of environmental decline. They may even cause further environmental decline if they are not contrived and added to projects in a way that is 'ecology-centric'. In a nutshell, the processes of reinventing, redesigning, remaking and recovering humanity's built environment need to be carried out *holistically*, and working at all geographic scales of reference.

ECOLOGICAL DESIGN IS NOT STATIC – CONSTANT ENVIRONMENTAL MONITORING, EVALUATION AND RESPONSE

Is Ecological Design a process with static outcomes? No effective design process can develop and grow effectively without carefully gathered and compiled feedback. Ecological Design requires the subsequent continuous real-time monitoring, assessing, measuring, comprehending, evaluating, predicting and mitigating of the impacts of humanity's interventions and of its designed system at each step of its materials' logistics and their route in their life-cycle on the global environment and on human society. The effort required is not static, but dynamic and constantly changing. Current technology and 'smart systems' based on digital technology can be used to monitor in real time and enable responses in real time to changes in urban conditions.

SUMMARY OF KEY ISSUES IN ECOLOGICAL DESIGN

Ecological Design is defined as an approach to problem-solving that provides the means and methods to resolve the physical environment aspects of the **sustainability equation**. Its ultimate goal is to transform our human-made world to provide a resilient, durable and sustainable future for all life on Earth, not just humanity. It is a process that involves redesigning, reinventing, remaking and repurposing our human-made world to be synergistically aligned with and supportive of Nature.

The implementation of Ecological Design involves setting and meeting performance objectives as far as practically achievable for as many factors as possible. Addressing all the multifarious and interacting considerations makes Ecological Design significantly more onerous and more complex than the conventional and traditional design approaches. Detailed and holistic ecological analyses and evaluations must lead the way. A key principle of Ecological Design is **ecocentricity**, where designing needs to be guided and directed by the science of ecology in all aspects of the designed system, from its spatial placement in the ecosystems to its production and the logistics of its materials over its flow and life-cycle. These considerations are further elaborated in Chapter 3, where a further principle of '**ecomimesis**' is introduced as the basis we should adopt to redesign, reconfigure and reinvent our human-made world.

NOTES

1 Carnegie Mellon University. 'Economic Input-Output Life Cycle Assessment,' www.eiolca.net/.

2 Ernest, S. K., and Brown, J. H. 'Homeostasis and Compensation: The Role of Species and Resources in Ecosystem Stability.' *Ecology*, vol. 82, no. 8, 2001, pp. 2118–2132.

3 Selge, S., Fischer, A., and van der Wal, R. 'Public and Professional Views on Invasive Non-Native Species – a Qualitative Social Scientific Investigation.' *Biological Conservation*, vol. 144, no. 12, 2011, pp. 3089–3097.

4 Simberloff, D. 'Confronting Introduced Species: A Form of Xenophobia?' *Biological Invasions*, vol. 5, no. 3, 2003, pp. 179–192.

5 Flynn, K. J., et al. 'Ocean Acidification with (De)Eutrophication Will Alter Future Phytoplankton Growth and Succession.' *Proceedings of the Royal Society of London B: Biological Sciences*, vol. 282, no. 1804, 2015, http://rspb.royalsocietypublishing.org/content/282/1804/20142604.

6 Von Wong, B. '4100lbs of E-Waste Resurrected. Trillions to Go.' *Von Wong Blog*, 19 Apr. 2018, blog.vonwong.com/dell/.

7 Baldé, C. P., Forti, V., Gray, V., Kuehr, R., and Stegmann, P. *The Global E-Waste Monitor 2017*, United Nations University, International Telecommunication Union and International Solid Waste Association, 2017.

8 'Biophilia' is a term first coined by the social psychologist and philosopher Erich Fromm in *The Anatomy of Human Destructiveness* (Holt, Rinehart and Winston, 1973). It was

later developed into a full theory by the eminent biologist Edward O. Wilson and eminent social psychologist Stephen Kellert.

9 Rittmann, B. E., and McCarty, P. L. *Environmental Biotechnology: Principles and Applications.* McGraw-Hill, 2001.

10 Chang, K. 'Hydrogen Cars Join Electric Models in Showrooms.' *New York Times,* 18 Nov. 2014.

11 Schiermeier, Q., et al. 'Energy Alternatives: Electricity without Carbon.' *Nature,* vol. 454, no. 7206, 2008, pp. 816–823.

12 Lovins, A. B. 'More Profit with Less Carbon.' *Scientific American,* vol. 293, no. 3, 2005, pp. 74–83.

FURTHER READING

Beatley, T. 2011. *Biophilic Cities: Integrating Nature into Urban Design and Planning.* Island Press, Washington, DC.

Beatley, T. 2014. *Blue Urbanism: Exploring Connections between Cities and Oceans.* Island Press, New York.

Carroll, B. & Turpin, T. 2009. *Environmental Impact Assessment Handbook.* Second edition. Thomas Telford, London.

Chartered Institute of Ecology and Environmental Management. 2018. *Guidelines on Ecological Impact Assessment in the UK and Ireland: Terrestrial, Freshwater, Coastal and Marine.* Chartered Institute of Ecology and Environmental Management, Winchester, UK. Available at https://cieem.net/wp-content/uploads/2019/02/Combined-EcIA-guidelines-2018-compressed.pdf.

Fuller, R. A. Irvine, K. N., Devine-Wright, P., Warren, P. H. & Gaston, K. J. 2007. Psychological Benefits of Greenspace Increase with Biodiversity. *Biology Letters,* 3, 390–394.

Kellert, S. R. 2005. *Building for Life: Designing and Understanding the Human–Nature Connection.* Island Press, Washington, DC.

Kellert, S. R. & Wilson, E. O., eds. 1993. *The Biophilia Hypothesis.* Island Press, Washington, DC.

Kellert, S. R., Heerwagen, J. & Mador, M. 2008. *Biophilic Design: The Theory, Science and Practice of Bringing Buildings to Life.* John Wiley & Sons, London.

Newman, P. & Jennings, I. 2008. *Cities as Sustainable Ecosystems: Principles and Practices.* Island Press, New York.

Nicholson-Lord, D. 2003. *Green Cities and Why We Need Them.* New Economics Foundation, London.

Ulrich R. S., Simons, R., Losito, B. D., Fiorito, E., Miles, M. & Zelson, M. 1991. Stress Recovery during Exposure to Natural and Urban Environments. *Journal of Experimental Psychology,* 11, 201–230.

3

Reinventing the built environment by 'ecomimicry'

THE BIOMIMICRY APPROACH IN THE REINVENTION OF THE HUMAN-MADE WORLD

How can we further address Ecological Design in an ecocentric approach? It has just been argued that Ecological Design must centre on effective biointegration, as the fused converging of our human-made world, and in particular its built environment, with Nature. The central issue is, then, how this is carried out? Proposed here is a conceptual and practical step in achieving this (as introduced in Chapter 1) through the expansion of the technique and approach of *biomimicry* in an ecocentric approach to the mimicry of 'ecosystems', referred to here as '**ecomimesis**'. This is the other key principle in Ecological Design.

A considerable body of research and endeavour has focused on biomimicry (earlier referred to as 'bionics', 'biomechanics' etc.), both within Nature and in relation to humanity's emulation of Nature. To elaborate on the brief definition given in Chapter 1, biomimicry *as applied to human design* is a nature-centric approach for designing based on emulating some of the evolved complexities seen in nature that 'solve' particular problems in efficient ways.[1,2] Fields of endeavour that are essentially extensions of biomimicry include bionics, biomechanics and bioengineering that involves the mimicry and imitating of biological systems. An example of this is the end window of the Crystal Palace in London, which was so large that engineers were struggling as to how to make it stable. The eventually successful design was modelled on the vein reticulation pattern of a giant leaf of the South American water-lily *Victoria amazonica*, after the designer Joseph Paxton reputedly noted that such a leaf at Kew Gardens could support the weight of a baby.

The premise for biomimicry is that humanity will most readily and rapidly find effective solutions to the issues and problems that humanity is currently confronting by examining the form and function of the complex, intricate and extensive biological systems in the natural world. The rationale for this is that Nature's world of living organisms and biological systems has undergone millions of years of development under the selective process of natural evolution. The end result is a complexity of interrelationships and functionalities between well-adapted species – a complexity that simply cannot be readily recreated from scratch. The natural world comprises systems, structures and processes that have been tried and tested over geologic timescales of natural selection and evolution.[3] To try to repeat that process of development and refinement in the short timescales of human existence would be, in most cases, impossible.

Taking a design approach based on biomimicry begins with first defining a specific problem or issue for which humanity is seeking a solution. Problem definition should then be followed by a process of defining the ecological, social and political context of the problem or issue. Following this is the search and identification of suitable biological living systems or organisms or processes from nature that may offer the potential and likelihood of solutions to the defined problem through mimicry.

This is followed by the extraction and derivation of the biological principles from the biological living systems or organisms or processes which can be applied to create solutions that can address the earlier-defined problem. Finally, in the process of design based on biomimicry, the derived biological principles are applied by the design to create a **prototype bio-mimicked solution** to solve the earlier-defined problem that in a rigorous approach needs to be tested before adoption by society.

Biomimicry as applied to Ecological Design is, then, an ecocentric approach based on the premise that in seeking to truly to start repairing Planet Earth and to avoid further harm, we need to become like Nature. We need to recreate our human world not only by borrowing elements from the library of Nature to aid our designs, as in the practice of biomimicry, but also by ensuring that the characteristics and functionalities of whole '**ecosystems**' are faithfully possessed by our human-made world. The term suggested here for this concept is *ecomimesis*. To advance this proposition, we must first articulate the meaning and nature of the term 'ecosystem'.

DIFFERENT DEFINITIONS OF ECOSYSTEMS

The intention is to derive the attributes of ecosystems for the purpose of Ecological Design. Ecologists use the term 'ecosystem' to refer to a particular identifiable system that we can find somewhere on Earth, and as a conceptual construct.[4] The difference between these needs to be elaborated here. Both are crucially important to the effective practice of Ecological Design.

The ecosystem as a specific unit in place

The term 'ecosystem' is often used to refer to a holistic system of dynamically interacting biotic and abiotic components over any defined region of Planet Earth. Both particular components and the natural forces influencing them, such as climate of the locality and the external disturbances of neighbouring ecosystems, can be defined. Generally, the ultimate source of energy is the sun, though ecosystems based on other energy sources exist in many parts of the Planet, on land and in the oceans.[5]

Ecologists often refer to ecosystems found on Earth with descriptors indicating varying degrees of 'naturalness'.[6] The way that humanity has destroyed or 'conquered' Nature is seen by many as setting humans apart from Nature, because the scale and rates of change affected by humankind no longer have any parallel or counterpart on Earth. Hence an ecosystem which appears to have arisen through processes not affected or little affected by humankind are generally considered 'natural', and those totally human-made (such as a lawn) are unnatural or 'artificial'. The most natural communities are also generally associated with the rarest and most specialised wildlife that has often evolved without anthropogenic interventions, frequently for the longest time.

In between these extremes, the term 'semi-natural' is sometimes used – e.g. for long-term cultural ecosystems that may owe their very existence and persistence to humankind and yet are hardly inhabited by humans and have extensive and often specialist communities of wildlife (e.g. heathland).

It should be immediately apparent that 'naturalness', though a useful concept and used in this book to refer to those systems with the least evidence of the influence of humans, can never be considered absolute. To begin with, even the most remote terrestrial ecosystems on Earth are impacted by anthropogenic changes such as climate change, air pollution and water pollution. Plastic bags were found at the bottom of the Mariana Trench (the deepest point in the ocean) as well as on deserted islands in the middle of the Pacific. Secondly, and perhaps more of a philosophical challenge, human beings, if considered the product of evolutionary process, are indeed part of Nature, hence any vestige of our presence, even the utterly abiotic artefacts, can accordingly be considered in some way 'natural'.

The ecosystem as a concept

The term '**ecosystem**' is also often used as a conceptual non-specific theoretical construct, such as in Systems Ecology. In this definition, ecosystems are simply any combinations of a variety of biotic constituents and abiotic constituents such as minerals and water interacting together to create a dynamic inter-dependent system with associated energy fluxes to produce a combination of form and function that, while variable, can be characterised.[7] For example, particular elements of ecosystems can be identified as a class of organisms operating at a particular trophic level.[8] While there can be considerable interaction and substitution of roles between levels in most terrestrial ecosystems, we have the following identifiable trophic-level categories, which aid ecosystemic analysis and understanding:

Primary producers, such as plants and algae, get their energy from the sun through the process of photosynthesis, combining sunlight, carbon dioxide and water to create organic tissue (or chemosynthesis for ecosystems that are based on a chemical energy source, such as at hydrothermal vents in the ocean). This provides the nutrient and energy base for the entire ecosystem.

Consumers, including all animals, cannot produce their own energy, so consume other nutrient-rich organisms, either primary producers or consumers lower in the food chain.

Decomposers, such as fungi and bacteria, get their energy from breaking down organic matter, getting their nutrients from previously living organisms. They return the rest to the soil as organic compounds, ready to be recycled back into the system.

The dynamic interactions between components result in additional emergent characteristics that cannot be found in the individual components; that is to say, *the properties of any ecosystem are larger than the sum of its parts.* Ecologists use this conceptual construct as the basis for theoretical predictions of the behaviour of Nature and for explaining measured natural phenomena.

In making these studies and predictions, 'Systems Ecologists' apply the principles of 'General Systems Theory' – a framework of thought dating back to the work of the Austrian biologist Ludwig von Bertalanffy in the early 20th century.[9] This framework, elaborating concepts and principles taken from thermodynamics, is used to set out principles, common elements and features of interacting components of the ecosystems and provide a rigorous means of addressing, analysing and describing them. The approach is holistic and contrasts with what is often called the reductionist approach, in which attempts are made to isolate particular elements of influences and determine their roles piecemeal.

ECOMIMICRY AS DESIGNING BASED ON THE ECOSYSTEM CONCEPT

How can we implement ecomimicry? The term 'ecomimicry' is used, then, to refer to the creation of a system in the human-made world that displays all of the key **attributes** and possesses the key **functionalities** of natural ecosystems, one of which is the seamless integration with existing natural ecosystems occurring in place and time.

To apply an ecomimetic approach, there is a need to identify the normative *attributes* of natural and semi-natural ecosystems. Once accomplished, Ecological Design is better able to understand and apply the general *functionalities* of ecosystems.

Following the listing and cross-referencing of these attributes and functionalities as normative (i.e. 'how things should or ought to be'), design based on 'ecomimesis' can be focused on creating the **best possible human-made system able to interface and function with the natural and semi-natural ecosystems.** The emergent properties (properties that can only be assessed when the ecosystem functions as a whole) can then be assessed, and if necessary, the design modified to try to optimise the properties and resultant functionalities.

IDENTIFYING THE KEY ECOSYSTEM ATTRIBUTES AND FUNCTIONALITIES TO EMULATE AND REPLICATE

What are the key attributes and functionalities of natural ecosystems that we might emulate in the rebuilding of our human-made world in implementing ecomimesis in Ecological Design? These also apply, in modified ways to semi-natural and full man-made ecosystems.

In the ecological sciences, there is an entire and burgeoning literature on the attributes of ecosystems. Those listed below include key examples of these attributes portrayed as simply as possible in a way that can be readily conceptualised, and point out how they might apply to the process of ecological and ecomimetic design, especially of the urban realm and the built environment.

Renewable energy base

Clearly, most terrestrial ecosystems acquire their energy from the sun as a renewable source and efficiently process that energy through complex trophic layers, an efficiency that generally increases as ecosystems head towards maturity.

The parallel for the built environment is obvious. Due to the potentially devastating consequences of continued climate change, we simply have to end our dependence on the use of polluting fossil fuels and look to systems that generate and conserve energy as close to the point of need and use as possible and see the biointegrated urban ecosystem as evolving towards ever greater efficiency in this regard.

Materials recycling

Mature natural ecosystems are very efficient at recycling their biotic and abiotic components, nutrients and energy. A good example is the tropical rainforest, in which nutrients are so well cycled by the various trophic levels that the soil itself is virtually barren, and if the tree cover is removed, soon disintegrates.[10]

The parallel for the creation of urban realm is to embrace throughout the concept of the 'closed loop' of both biotic and abiotic materials, including reuse and recycling and replenishment of materials, including their benign replenishing and reintegration back into the environment.

Ecological succession

Ecosystems change through a progression of species assemblages, new species joining and some existing species leaving the system over time from its inception, generally with a steady increase in overall biomass of the system. This is known as '**ecological succession**'. In emulating this attribute, human artefacts and urban realm should be viewed as dynamic ecological systems that are *managed* as much as *master-planned*. For the most resilient and functional hybrid urban ecosystems to develop, allowance needs to be made in space, time and management for the processes of succession to occur. Provision of links for species to colonise newly created habitats and contribute to succession is essential. In some cases where species will not be able to colonise naturally, succession may need to be 'arrested by management' at a given functional stage to optimise required ecosystem service delivery, but then the products of that management need to be fed back into the same system, fuelling it.

Ecosystem homeostasis

Ecosystems are always to varying extents in a state of flux. Their organisms, numbers and relationships change in a processes of succession, local extinction and recolonisation. The complex interactions between the species present and their physical environment control the patterns, rates and limits of this succession. Cumulative changes in many smaller ecological systems can add up to significant impacts on larger ecological systems.

Even the most stable ecological systems change when affected by cyclic or catastrophic disturbances. However, over evolutionary time, natural selection has favoured those ecosystems that can recover to their former (or a similar) state of equilibrium, much as a warm-blooded animal can generally control its own body temperature.[11]

When we apply these characteristics to our thinking in the creation of the built environment, we need to consider cybernetic systems of monitoring and feedback that enable the created system to respond to perturbations in real time, whether the perturbation is due to anthropogenic factors or wider environmental change.

The extensive and complex monitoring involved ideally requires a continuous biosphere-wide inventorying, remote-sensing-and-response system of the state of the Planet's ecosystems, of its biogeochemical cycles and of the character of human acts and activities. This needs to link seamlessly to corrective mechanisms that result in near instantaneous preventive, corrective and restorative action.

Ecosystem interconnectivity

While scientists and theoreticians can define separate ecosystems, unless they are in a laboratory test-tube, they are not in fact separate at all, but linked in various ways to a variety of other ecosystems, both near and sometimes far away. The influx of species and genetic variety are important factors in maintaining the vitality and adaptability of ecosystems in Nature. Furthermore, some species depend on a variety of different ecosystems for the completion of their life-cycles.

The ecomimetic design of our world therefore needs to incorporate a 'nexus' of interconnected designed and incorporated natural and semi-natural habitats integrated with the wider ecological infrastructure (referred to as 'augmentation of the built systems' in Chapter 5). Careful science needs to be applied to the design of linkages, tailoring them to the researched requirements of species and making them as multifunctional as possible (see Chapters 4 and 5).

But beyond the connections of habitats there needs to be connections between the materiality of the constructed urban realm and the Planet's biogeochemical cycles.

Ecological 'integrity' (based notably on species diversity)

Ecological integrity is a complicated concept about which much has been written.[12] In simple terms, for an ecosystem to be healthy and to provide ecosystem services (see below), it must have the full complement of components (biotic and abiotic) at the appropriate levels and in the optimum interactive balance of interactions.[13,14,15] Clearly there is leeway, but due to homeostatic mechanisms, ecosystems (see later in text) generally return to the functional optimum, after perturbations, for the given set of environmental conditions. As a generalisation, the greater the species diversity in a given type of ecosystem, the greater the resilience of that ecosystem to environmental changes and shocks. This is partly a function of the greater *redundancy* in the system: when one component fails or becomes extinct another can assume its functional role and maintain the integrity of the system as a whole.

When designing our artefacts and built environments ecomimetically the aim is to ensure supply, or the provision of functional routes for addition or natural colonisation, of the full integrated complement and balance of both biotic and abiotic constituents (e.g. the physical and the living organic constituents, including all locally appropriate species and varieties of organism). The goal becomes one of nurturing the development and persistence of the full complexity seen in contextual ecosystems of the region for which we are designing.

In other words, we need to develop to maximise biodiversity at all levels – habitat, species, subspecies, and indeed individuals of species and ecological processes in the urban realm – that is locally appropriate to the ecological reference context (see Chapter 4, where the topic of ecological targets for a given project is discussed further).

Ecosystem functioning and provision of ecosystem services

Ecosystems 'function' in that they provide a basis for the sustained maintenance of all life. These functionalities include the furnishing of life with an equable climate, healthy air, healthy water, healthy land, healthy food, shelter and other materials, and sustained energy. When the life that is benefiting from the functionalities of ecosystems is human, the benefits accruing are generally referred to as 'ecosystem services'. However, this is a highly anthropocentric definition, and we need to reconsider the term 'ecosystem services' as applying to all lifeforms on the Planet.

In the design of any system or piece of urban realm following ecomimetic principles, added value results from the fact that the resultant systems **function** to provide such ecosystem services. The topic of ecosystem services is further elaborated in Chapters 5 and 6, but in summary, designed biointegrated elements in human artefacts and urban areas can contribute significantly to climate control, water management, water purification, production of food and fuel, and nutrient cycling, among others. The integration of this functionality into any designed system makes it less dependent on the provision of the services by other ecosystems, near and far.

There is not yet a universal popular demand by humanity to ensure that the design of our artefacts and built environment sustain Nature's ability to provide ecosystem services, but that may change with the burgeoning of the human population, exponential growth of the built environment and the ongoing decline in the capacity of supporting ecosystems 'outside' of our designed realms to provide all that we require for healthy and productive life.

OTHER ECOSYSTEM ATTRIBUTES

Besides the above attributes, a variety in climatic conditions lead to the establishment of different biomes. The practical approach to emulating ecosystem response to climate in Ecological Design of the urban realm is to create a built environment that responds to the climate of its locality. This 'bioclimatic' design approach would create built systems that respond in different configurations to different climatic conditions.

A PERFORMANCE-BASED FRAMEWORK

It is argued here that the above set of attributes and emergent functionalities of ecosystems provide the crucial general performance-based framework for the sustainable reinvention of our human-made world. It should be added that it is very difficult (because of the emergent properties of ecosystems) to be fully prescriptive about the functionalities that will obtain from systems designed on the principles of ecomimicry. However, performance-based targets can still be set for all functionalities, and the designed systems created then monitored and modified over time to achieve them.

THE CREATION OF THE BUILT ENVIRONMENT AS MULTI-BRID LIVING SYSTEMS

Our existing cities and other urban areas (as stated in Chapter 1), as humanity's greatest and largest global artefacts, do not currently as a rule integrate benignly with nature. Instead they generally pollute, produce net waste and consume non-renewable resources. Because of the implications of these effects on the survival of most life on Earth, our existing and new built environments need to be reinvented and transformed as far as we can into 'novel constructed ecosystems' – multi-brid 'living' systems formed from a seamless interlinking of synthetic and semi-synthetic human constructions with natural systems – systems that serve as integral parts of wider Nature, and not separated from it.

We can refer to the concept of 'techno-biologically constructed ecosystems' as hybrid composite systems, rather like the 'eco-cyborgs' described in popular science fiction. Clearly, this goes beyond just having green spaces in cities, even those with significant biodiversity. It refers to the remaking of the built environment as a *whole multi-brid living system*, not as one with dispersed green areas and patches within. The goal is no less than removing the distinction between urban and non-urban areas, and indeed between the synthetic and the natural, and replacing it with a fused, seamless, synergistic and tightly biointegrated construct.

KEY PRINCIPLES IN IMPLEMENTING ECOMIMESIS

Key principles to direct the overall approach need to be borne in mind here when embarking on an ecomimetic path or reinvention of our world.

Firstly, it should be *properly purposed*. Applying ecomimicry to design is not about applying and replicating a set of attributes to secure a comfortable life for humankind. It needs to be philosophically and morally purposed, aimed at resolving the key problems that confront humanity today, and resolving the multiple aspects inherent in the 'sustainability equation' (see Chapter 1). This in turn means addressing existing environmental and social damage, preventing new damage, becoming as ecologically effective as possible and contributing to human civilisation.

Secondly it should be *fully contextual*. Any hybrid ecosystem made by humans must not just be functional in its own right, but integrate seamlessly and both structurally and functionally with the systems nearby and in the wider region. This may entail the creation of bio-engineered replicates of naturally occurring ecosystems within the design, integrated seamlessly with the rest of the urban infrastructure. This relationship needs to be effective not just locally, but at all scales of interaction with its bioregion, biomes and across the entire biosphere. The comprehensiveness of this goal, then, differentiates Ecological Design from conventional design.

Thirdly it should be *all-encompassing*. For example, it should not stop at the boundaries of the created artefact or urban area, but address all steps in the logistics flow of the components used in making the artefact or place over their life-cycles. It needs to be further applied to the built environment's technical and operational systems, and to the social political systems of the humans that will use the artefact, influencing minds towards a more benign and sustainable relationship with Nature.

Fourthly it should seek to *optimise benefits for both people and Nature*. Clearly, by optimising delivery of ecosystem services, people will benefit. But by carefully integrating technology with Nature, we can potentially arrive at what might be considered 'Nature-augmented' or Nature-enhanced solutions that increase the innate resilience or productivity of natural or semi-natural ecosystems, e.g. by providing additional refuges for the species of which they are composed.

REINVENTING THE BUILT ENVIRONMENT BY 'ECOMIMICRY'

What is the current status of this endeavour? Various prototype solutions have been implemented singularly or severally in a variety of built systems and are hence being tested for adoption by society. We have yet, however, to see these being integrated into a complete and holistically designed single system. This is, then, the key current challenge for humanity in ecomimetic Ecological Design as the other key pillar in Ecological Design.

Successfully accomplishing ecomimesis in adopting the 'ecosystem' concept as the model for emulation and replication for humanity's entire urban conurbations means that these conurbations cease to be isolated and distanced from naturally occurring ecosystems and cease to be parasitically dependent on nature outside the built environment's artificially defined boundaries for the crucial inputs.

The ideal reinvented and reconstituted built environment becomes an ecologically effective and functioning integral part of the biosphere. It is should be noted that some of the extensive ecosystem damage that humans have already inflicted on Earth may have become irreversible. We cannot as yet, for example, bring extinct species back into existence. Accordingly, we may need to accept that our endeavours to repair and restore systems will in some cases be met with localised and incremental success despite humanity's technological prowess and ability to innovate. Moreover, human socio-political systems are unpredictable and currently very far from permitting a global adoption of ecomimetic principles. However, we would argue that such limitations should not inhibit the implementation of as full as possible an ecomimetic approach wherever possible.

NOTES

1 Benyus, J. M. *Biomimicry: Innovation Inspired by Nature*. Perennial, 2009.

2 Pawlyn, M. *Biomimicry in Architecture*. RIBA Publishing, 2011.

3 Darwin, C. *On the Origin of Species*. John Murray, 1859.

4 Cain, M. L., et al. *Ecology*. Sinauer Associates, Inc. Publishers, 2014 (Ch. 1).

5 Wirsen, C. O., et al. 'Chemosynthetic Microbial Activity at Mid-Atlantic Ridge Hydrothermal Vent Sites.' *Journal of Geophysical Research*, vol. 98, no. B6, 10 June 1993, p. 9693. doi:10.1029/92jb01556.

6 Molla, A. S. 'Natural, Semi-natural, and Artificial Ecosystems.' *The Daily Star*, 29 Nov. 2008.

7 Bunker, A. F. 'Computations of Surface Energy Flux and Annual Air–Sea Interaction Cycles of the North Atlantic Ocean.' *Monthly Weather Review*, vol. 104, no. 9, 1976, pp. 1122–1140. doi:10.1175/1520–0493(1976)1042.0.co;2.

8 Pimm, S. L., and Lawton, J. H. 'Number of Trophic Levels in Ecological Communities.' *Nature*, vol. 268, no. 5618, 17 May 1977, pp. 329–331. doi:10.1038/268329a0.

9 Von Bertalanffy, L. *General System Theory: Foundations, Development, Applications*. Braziller, 1980.

10 WildMadagascar.org. 'Why Rainforest Soils Are Generally Poor for Agriculture.' 2019, www.wildmadagascar.org/overview/rainforests2.html.

11 Ernest, S. K., and Brown, J. H. 'Homeostasis and Compensation: The Role of Species and Resources in Ecosystem Stability.' *Ecology*, vol. 82, no. 8, 2001, pp. 2118–2132. doi: 10.2307/2680220.

12 Ocean Health Index. 'Ecological Integrity.' 2019, www.oceanhealthindex.org/methodology/components/ecological-integrity.

13 Costanza, R., et al. *Ecosystem Health: New Goals for Environmental Management*. Island Press, 1992.

14 Lu, Y., et al. 'Ecosystem Health towards Sustainability.' *Ecosystem Health and Sustainability*, vol. 1, no. 1, 2015, p. 2. http://dx.doi.org/10.1890/EHS14-0013.1

15 Rapport, D. 'Assessing Ecosystem Health.' *Trends in Ecology & Evolution*, vol. 13, no. 10, 10 Oct. 1998, pp. 397–402. doi:10.1016/s0169-5347(98)01449–9.

FURTHER READING

Grant, G. 2012. *Ecosystem Services Come to Town: Greening Cities by Working with Nature.* Wiley-Blackwell, London.

Naeem, S. et al. 1999. *Biodiversity and Ecosystem Functioning: Maintaining Life Support Processes.* Issues in Ecology, 4. Ecological Society of America, Washington, DC.

Wells, M. J., Timmer, F. & Carr, A. 2011. Understanding Drivers and Setting Targets for Biodiversity in Urban Green Design. In: K. Yeang & A. Spector, eds, *Green Design from Theory to Practice.* Black Dog Publishing, London, 13–32.

Wells, M. J. & Yeang, K. 2010. *Biodiversity Targets as the Basis for Green Design. Architectural Design Magazine* Exuberance Profile No. 204: New Virtuosity in Contemporary Architecture, 130–133.

TUCK

4

Ecological Design as the biointegration of a set of 'infrastructures'

The 'quatrobrid' constructed ecosystem

AN INTEGRATIVE FRAMEWORK FOR ECOLOGICAL DESIGN

What are the practical means for implementing Ecological Design? While ecocentricity means adopting the science of ecology as the guiding principles for achieving the multiple tasks mentioned earlier, and whereas ecomimesis provides a model of attributes for creating an ecocentric reinvention of the human-made world and built environment, we need a framework for practical implementation of Ecological Design.

In Ecological Design, we need to integrate the natural world with the built environment, including all components of Nature from biogeochemical cycles to flora and fauna, and all built components of human society from managed water systems to artefacts to societal systems.

In providing a framework for the synthesis of these into a designed system, it is useful to think each of these as a support system in terms of what is referred to here as an '**infrastructure**'. Ecological Design is, then, combining the balanced, seamless and ecologically informed blend of these constituents into a biointegrated composite 'constructed ecosystem' based on ecomimetic principles. Ecological Design brings together the four factors as infrastructures synergistically biointegrated into a composite four-part system, referred to here as a '**quatrobrid**' constructed ecosystem consisting of the following components.

- **Nature-based infrastructure**

Commonly referred to as 'green' infrastructure, this is the environmental context for all the other infrastructures, as it is the Planet's life support system and consists of the Planet's ecosystems and biogeochemical cycles.[1,2,3,4,5,6,7] The term applies whether the infrastructure is naturally developed or designed and planted.

- **Technological and engineering infrastructure**

The technological and engineering infrastructure comprises the physical constructs that humanity designs and manufactures. It includes all of humanity's service-providing artefacts, structures and technologies. These consist both of the 'unenclosed' urban utilities and the 'enclosed' internal mechanical/electrical/information technology servicing systems within them. While technological and engineering infrastructure clearly includes the technological components of the hydrological infrastructure, these are considered separately as part of hydrological infrastructure.

- **Hydrological infrastructure**

'Hydrological infrastructure' refers to the parts of both Nature-based infrastructure and our technological and engineering infrastructure that are particularly associated with water and water management.[8,9,10] Put another way, it is humanity's hydrological and water management and reticulation systems acting in unison with the natural cycles of the hydrosphere. The reason why the water-related infrastructure is often treated separately is because of the paramount importance of water to life. The natural components of hydrological infrastructure include natural watercourses and waterbodies and groundwater. The human-made components include drainage networks, water storage features, flood defences, pumps, sumps and sewage networks.

- **Human socio-economic-political infrastructure**

The terms 'socio-economic-political infrastructure' or 'anthropological infrastructure' are used here to refer to the complex societal social-economic-political-institutional systems and communities of humanity. Humans, while part of Nature, have undergone evolution based on natural selection to arrive at a state of cultural evolution whereby ecologies of societal interaction, and indeed thoughts, have enormous effects on the Planet. Accordingly, socio-economic-political infrastructure must be considered as an infrastructure in its own right.

It is important to note that these categorisations are not mutually exclusive and are used largely to facilitate designing to enable the endeavour of Ecological Design to focus on a group of particular aspects of the fabric of our human-made world and Nature united by a common element.

RATIONALE FOR AN INFRASTRUCTURE-BASED APPROACH TO ECOLOGICAL DESIGN

Each of these 'infrastructures' can be considered an underlying *scaffold* or *backbone* in a network of sub-systems from which the holistic functions of ecomimetic design emerge. These infrastructures in totality encompass all the key factors that make up human society and its built environment and Nature. Ecological Design is, then, the effective biointegration of this set of infrastructures into a whole system where an infrastructure is an organisation of entities, whether already in existence in Nature (as in the Nature-based infrastructure, or the hydrological infrastructure), or is a societal construct, being its anthropological infrastructure and technological infrastructure. The concept of different types of infrastructure has been introduced in previous chapters. It is worth recapping here in more detail before explaining why Ecological Design must operate at the level of infrastructures.

The rationale for the adoption of an 'infrastructural' approach is that Ecological Design, to be properly effective, needs to be carried out in a system-wide macro scale and not be undertaken in an incremental or piecemeal way, such as at the scale of just individual human constructs (such as buildings) or even multiple constructs.

Ecological Design at micro scales of working, at its best, can achieve only local and incremental environmental benefits. Even if we deliver buildings with state-of-the-art energy-saving systems and the most ecologically effective systems, such endeavours are likely to remain ineffectual in reversing the already advanced biosphere-wide condition of environmental impairment unless undertaken as part of a wider change in approach and delivery.

Moreover, if individually designed green buildings and structures are connected to an infrastructure that is not ecologically benign, which impairs Nature, then no matter how extensively 'sustainable' or ecologically effective they are, the entire system continues to contribute to the overall adverse environmental impacts of environmentally defective human operation and urbanism. A simple example of this might be where the buildings are connected to an electricity grid whose source of electricity comes from the burning of fossil fuels. In this instance, the buildings and structures connected to the grid cannot be regarded as forming an 'ecologically effective' system.

Seeking to achieve sustainability through the incremental delivery of individual green buildings without addressing our human-made world holistically is somewhat like recycling at home. Such efforts can make the individual or the local communities feel that they have done something right, but in reality, the overall beneficial environmental impact is insignificant if they are not part of an overall effective system that strives to eliminate the very concept of waste.

When the infrastructures of our conurbations are ecologically effective and green, the buildings and other components that are serviced by them can also become ecologically effective. Viewing the components of our human-made world in terms of infrastructures also helps us to ensure that the all of the effects of our designed systems and their components are considered at every stage in the life-cycles of their components. In particular, the ecological effectiveness of technological and hydrological infrastructures needs to be optimised through consideration of all environmental impacts from the source of supply for each infrastructure, through their distribution systems, to their 'end of use'. Furthermore, as has been emphasised earlier, every act of Ecological Design needs to address, as far as possible, the repair of past environmental damage, and to do this effectively, we need to work at all scales, considering all interdependencies – that is to say, again, working with entire infrastructures rather than localised individual components.

PHYSICALLY AND SYSTEMICALLY INTEGRATING THE SET OF INFRASTRUCTURE INTO A DESIGNED SYSTEM

Although each of these infrastructures can be defined to some extent independently, they are also clearly interdependent. They need to be assembled, inter-woven, unified, enmeshed and biointegrated into ecologically designed systems, the interactions and synergies between which can be envisaged as combining and interacting in a multi-dimensional '**matrix of infrastructures**'.

Each infrastructure must be designed to emulate and replicate the normative attributes of ecosystems to be in synergy with wider Nature. The whole biointegrated **matrix of infrastructures**, if you like, becomes a designed **'quatrobrid**' (a term that may be preferable to the binary 'hybrid') techno-natural constructed ecosystem. This integration needs to operate at all scales of operation and influences, from the macro to the meso to the micro and nano scales.

The biointegration and making of the infrastructure matrix must be designed to operate within the thresholds of the resilience and carrying capacities of Nature's ecosystems and the biosphere's biogeochemical cycles, foster the conservation of its natural resources, and where achievable, contribute positively and beneficially to the restoration and integrity of Nature.

RECONCEPTUALISING THE ECOLOGICAL DESIGN PROCESS

In summary, to effect proper Ecological Design, the full set of infrastructures in Chapter 5-8 in the design framework are the crucial elements that need to be brought together and biointegrated to make up the hybrid designed system. Essentially, we can then conceive the practical process of Ecological Design as the seamless, benign and synergistic biointegration of the natural, hydrological, technological and anthropological infrastructures. We can refer to these not as creating a 'hybrid' constructed ecosystem, but a 'multi-brid' system, in this instance, there being four factors, a 'quatrobrid' constructed ecosystem.

The biointegration and the design of each infrastructure and the assembly of the infrastructures into a combined synergistically integrated multi-dimensional interactive system must emulate and replicate holistically the normative attributes of ecosystems through **ecomimicry**, thereby functioning as a new type of ecosystem – a human-made ecosystem in harmony with Nature. The resultant system must work within the thresholds of the resilience and carrying capacities of Nature's ecosystems and the biosphere's biogeochemical cycles, and the conservation of its natural resources, and where achievable, contribute positively and beneficially to Nature. Accomplishing this integration effectively is the core challenge of Ecological Design.

NOTES

1 Andersson, E., et al. 'Reconnecting Cities to the Biosphere: Stewardship of Green Infrastructure and Urban Ecosystem Services.' *Ambio*, vol. 43, no. 4, 17 Apr. 2014, pp. 445–453. doi:10.1007/s13280-014–0506-y.
2 Edwards, P. E. T., et al. 'Investing in Nature: Restoring Coastal Habitat Blue Infrastructure and Green Job Creation.' *Marine Policy*, vol. 38, 12 June 2012, pp. 65–71. doi:10.1016/j.marpol.2012.05.020.
3 Foster, J., et al. *The Value of Green Infrastructure for Urban Climate Adaptation.* Center on Clean Air Policy, 2011.
4 Hansen, R., and Pauleit, S. 'From Multifunctionality to Multiple Ecosystem Services? A Conceptual Framework for Multifunctionality in Green Infrastructure Planning for Urban Areas.' *Ambio*, vol. 43, no. 4, 17 Apr. 2014, pp. 516–529. doi:10.1007/s13280-014–0510-2.
5 Kazmierczak, A., and Carter, J. *Adaptation to Climate Change Using Green and Blue Infrastructure: A Database of Case Studies.* University of Manchester, June 2010.
6 City of New York, Office of the Mayor. *NYC Green Infrastructure Plan: A Sustainable Strategy for Clean Waterways.* New York City Department of Environmental Protection, 2010.
7 Tzoulas, K., et al. 'Promoting Ecosystem and Human Health in Urban Areas Using Green Infrastructure: A Literature Review.' *Landscape and Urban Planning*, vol. 81, no. 3, 2007, pp. 167–178., doi:10.1016/j.landurbplan.2007.02.001.
8 .Karamouz, M. *Hydrology and Hydroclimatology: Principles and Applications.* CRC Press, 2013.
9 Margulis, S. *Introduction to Hydrology*, 4th ed., 2017, https://margulis-group.github.io/teaching/.
10 Dang, T. D., et al. 'Hydrological Alterations from Water Infrastructure Development in the Mekong Floodplains.' *Hydrological Processes*, vol. 30, no. 21, 29 Apr. 2016, pp. 3824–3838. doi:10.1002/hyp.10,894.

FURTHER READING

Construction Industry Research Association. 2015. *The SUDS Manual.* Construction Industry Research Association, London.

5

Nature-based infrastructure – Earth's 'life support system'

NATURE-BASED INFRASTRUCTURE

Nature-based infrastructure in its broadest sense is our life support system and is essentially synonymous with the **biosphere** – the thin 'envelope' that surrounds the Planet that extends out about 10,000 metres (6 miles) from the Earth's surface.[1] The biosphere stretches from the lower atmosphere, the hard surface of the Earth (lithosphere) and the hydrosphere (waterbodies, watercourses and aquifers), and contains all the forms of life that exist on Earth. Relatively recently, it has been found that life occurs deep underground in geological crevices. How and to what extent these elements of the global biota affect the rest of the biosphere is a matter of ongoing research. However, the tasks of repurposing, reinventing, redesigning, remaking and recovering our human-made world and biointegrating it seamlessly and synergistically into natural and semi-natural ecosystems must start with consideration of Nature-based infrastructure, regarded here as the most important and crucial design determinant and core framework into which other infrastructures must be integrated.

The Earth's living creatures, ultimately fuelled in most cases by the sun's energy, interact with the Earth's biogeochemical processes and cycles, forming a complex multi-dimensional web of lifeforms and processes that have many emergent properties greater than the sum of the parts.

As stated in Chapter 2, the Earth's biospheric ecosystems provide humankind and other forms of life with resources (goods) and services which are collectively referred to as **ecosystem services** (see later in this chapter). It is this concept of a *relationship with and service provision to humans* that leads to the use of the term **infrastructure** to describe the framework of ecosystems, *both natural and human-made*, considered and defined in terms of the services that they supply to humankind. Clearly, this means that Nature-based infrastructure (whether natural or designed) will be to a greater or lesser extent multifunctional and hence of multiple utility to humans and other species.

Nature-based infrastructure is referred to by some as 'Nature's utilities', analogous to 'servicing functions' in humanity's urban engineered utilities, although Nature-based infrastructure is considerably more complex and intricate.

NATURE-BASED INFRASTRUCTURE AND 'GREEN INFRASTRUCTURE'

Before delving further into types of ecosystem services, it will be useful to consider a way in which Nature's infrastructure is often referred to in urban planning around the world, namely as '**green infrastructure**' (GI).[2,3,4] Much has been written about the term in the academic literature, and it is clear that the concept of GI is still evolving and that definitions are often formulated to suit the needs of a given context. GI has even been defined as encompassing a combination of physical components of the landscape on the one hand and an abstract 'conceptual approach' to urban planning on the other. However, the term is perhaps most frequently used to refer to networks of green spaces dominated by flora and non-human fauna and which provide services to humankind, often in an urban context, but also applied in suburban and rural contexts (more will be said on this later in this chapter). One of the key services of green infrastructure to humankind in urban areas is the biophilic enhancement of the experience of movement of people through the urban (or degraded rural) fabric.

Green infrastructure is generally considered and analysed at all scales of analysis and intervention, from the macro to the micro. This is indeed essential to ensure proper and effective biointegration of the human-made and more natural systems in a given bioregion.

NATURE-BASED INFRASTRUCTURE AND ECOSYSTEM SERVICES

The Nature-based infrastructure freely provides 'ecosystem services', which are the services it provides without human intervention and which are essential to sustain all lifeforms on Earth, including humans.

There are a variety of systems by which ecosystem services have been classified, but the most generally accepted framework is provided by the Millennium Ecosystem Assessment in 2005, which classifies ecosystem services according to functional groups of supporting, provisioning, regulating and cultural services. Note that this system is essentially anthropocentric, and in fact all the services provided to humans are also services supplied to all other species with the same arguable right to exist on Earth.

These ecosystem services will now be considered briefly in turn.

Supporting services

These are supporting functions of ecosystems that are crucial for the production of all other ecosystem services. These differ from the other ecosystem services in that changes to them, and hence impacts on humans, are generally affected over much longer periods of time and/or indirectly. These supporting ecosystem services include:

Primary productivity and oxygen production – the production of organic matter through processes such photosynthesis, producing oxygen as a vital by-product, and chemosynthesis.[5]

Nutrient cycling within the natural environment (storage, cycling and maintenance of nutrients by organisms and their associated processes) – nutrients are recycled through the life-cycle of organisms as they die and decompose, releasing the nutrients into the neighbouring environment.[6,7,8,9]

Soil formation and retention.

Provision of habitats and maintenance of biological and genetic diversity.

Pollination.[10]

Provisioning services

Ecosystems provide materials that are used by humankind in a multitude of ways, including:

Food and fodder production (food and spice crops for people, animal feed crops, wild-harvested fauna crops).

Fibre and building materials (e.g. many natural fibres for clothing and wood for construction of tools and artefacts).

Fuel production (wood, dung and other biological materials that can produce energy via combustion or fermentation).

Natural medicines and biocides.

Genetic source material for the genetic diversification, and hence long-term adaptability and resilience, of other ecosystem products (e.g. crop improvements genes, test and assay organisms).

Freshwater (e.g. freshwater in lakes, rivers and streams, in frozen state, soil moisture) – see also regulating services.

Stored water, generally from impoundments along watercourses, with gravitational potential that can be exploited to produce hydroelectrically generated power.

Regulating services

These are the services arising from ecosystem processes that maintain the global cycles within the biosphere in functional balance. Examples include:

Climate and weather regulation, which is one of the crucial services of global ecosystems: global ecosystems act as a carbon 'sink', sequestering greenhouse gases.[11] Included within this is the regulation of the chemical composition of the oceans. The oceans are also the key global heat sinks. Locally, ecosystems can affect local climate, weather and microclimate very significantly.

Water regulation (including storm protection, flood attenuation, groundwater recharge, water cycling and global distribution of freshwater).

Erosion control.

Water purification through natural processes.

Waste decomposition, detoxification and recycling (e.g. absorption and filtration of chemicals by soils).[12]

Air quality regulation.

Wind attenuation.

Noise attenuation.

Biological control of pests and diseases (e.g. control by insects, birds, bats and other organisms).

Ecosystem disturbance regulation – ecological homeostasis.[13]

Protection against harmful cosmic radiation.

A crucially important example of the regulating service provided by nature-based infrastructure in our urban areas is to combat so-called 'urban heat islands'. The sun's radiation is absorbed by the predominantly dark surfaces of many urban areas and then at night is re-radiated as heat that raises the urban temperature substantially above that in the rural hinterland. The temperature rises in many large modern conurbations can be high enough to be life-threatening through heat stroke, especially when heat waves occur, when many thousands of premature human deaths may result. The temperatures also worsen the effects of other urban ills such as air pollution. Combating the urban heat island effect is an area of very active research. Vegetation can play an extremely important role, as long as it is in a good enough physiological condition (see Chapter 6 on hydrological infrastructure).

Anthropocentrically beneficial services

These are services that benefit humans that may have no immediate benefits for the ecosystems and which are the largely non-material benefits humans obtain from ecosystems. While largely non-material, however, these can have huge economic implications. These services include:

Biophilic responses in terms of physical and psychological health and well-being.
Educational processes and approaches.
Inspiration of art, culture, genius loci.
Social structural cohesion and knowledge systems (e.g. the relationship between natural resources and harvesting or hunting groups).
Spiritual and religious values.
Recreation and ecotourism.

It should be apparent from this summary that the ability of properly functioning ecosystems in the biosphere to furnish ecosystem services is the most crucial ecosystem attribute that humanity must preserve and within the limits of which humankind must operate. Nature's provision of ecosystem services is finite, as is the Earth's ability to furnish us with non-renewable natural material and energy resources.

UNDERSTANDING AND ANALYSING ECOSYSTEM SERVICES PROVISION

The assessing of Nature's ability to provide **ecosystem service**s is a complex undertaking. Analyses include:

Assessment of *ecosystem processes and qualities* such as competition, dominance, population balance, invasion, succession, patchiness, perturbation, variability, resilience, resistance and persistence.
Assessment of the *effects of scale* on the above processes and qualities, often using geographic information system analyses.[14] An example watershed analysis includes components at the scale of the entire watershed and interactions between watersheds down to individual components of a drainage network.
Analyses of the *material flows* through ecosystems – i.e. flows of materials and energy through the metabolic pathways within ecosystems, and the undertaking of 'ecological footprint' studies (see later in this chapter).

It should be noted that while the impacts we exert on Nature may be negative, e.g. resulting in the loss of key species and destruction of key habitats, they can also be positive, such as reinforcing the integrity of (especially degraded) ecosystems and their ability to deliver ecosystem services.

RELOOKING AT LIMITING PARAMETERS

We need to re-examine the limiting parameters of global ecosystems to determine at the outset how to design and operate our human-made world.

The designer must be aware of the components of natural infrastructure in every project and apply the precautionary principle, trying to ensure that the ecological integrity of affected ecosystems is maintained or enhanced.

Natural limiting parameters need to set limits on the type of designed systems, artefacts and built structures humanity should produce, and how these are produced, used and recovered. They should influence the material content, mode of manufacture and operation over system life-cycles from first use to final resorption into the environment.

DESIGNING NATURE'S INFRASTRUCTURE FOR OPTIMAL ECOSYSTEM SERVICES PROVISION

The ability of ecosystems to furnish humans and other species with ecosystem services needs to be sustained rather than undermined by our actions. The synergy in these ecosystem services is such that if we remove any of one of them, the others start to break down and eventually disappear. Their protection and restoration must be the mandatory requisite practice for all of humanity's acts within the global environment and for Ecological Design.

The poor ability of the built environment to provide significant ecosystems services is not because of the total absence of biotic constituents and green spaces. The inability generally results from a combination of several deficits:

The frequently limited areas/sizes of the biotic components and the enclosure of what green components there are by an inimical mass of inorganic and often relatively inert built structures dominated by concrete, masonry and steel.

The frequently impoverished biodiversity of the components, which in turn diminishes all ecosystem functions and desired properties such as resilience and stability.

The fragmentation of the biotic elements within the overall urban fabric and poor consideration of the optimum relative proximity to and connectivity with each other.

The limited three-dimensional extent of the green elements.

The lack of true synergistic biointegration with the built systems.

We will now consider each of these aspects.

THE NEED FOR SCALE

As we are now a predominantly urban species, our sheer concentration of numbers and needs places enormous pressures on remaining pockets of Nature inside urban areas. Around the world, many such pockets each day are converted into new built development. Such excess of development produces changes in the faunal species that survive in urban areas. Although remnant natural habitats and artificially created new semi-natural habitats supply many of the habitat requirements of the native species typical of the region, not all can colonise, and if they do, not all can persist. Many need much larger areas far from edge effects, or juxtaposition to other types of habitat to meet their full life-cycle requirements, for example.

In other words, when designing urban green infrastructure, we need to understand species differences. For example, the following classification of species is often discussed by urban ecologists:

'**Urban avoiders**' – species that avoid living in the natural habitats of cities because of a lack of necessary requirements.

'**Urban utilisers**' – species that can to varying extents utilise habitats inside urban areas that match the species requirements.

'**Urban dwellers**' – species which prior to the city development may not have existed in the area, but now specifically favour or are adapted to the habitats provided in many urban areas and/or which may favour close proximity of humans, sometimes referred to as '**synanthropic**' species.

One of the reasons why many species may be urban avoiders, or only occasional urban utilisers, is the lack of scale of suitable habitat. Any patches of suitable habitat in most cities, tend to be set in a sea of inimical habitat. If the entire ecosystem of a city changed to one of extensive and intensive biodiverse habitat coverage, who knows which species traditionally considered to be urban avoiders might colonise? In essence, the existing green areas in nearly all urban areas today are not of sufficient size/area nor sufficiently effective in design and composition to provide for a suitably biodiverse semi-natural, resilient and stable assemblage of species to underpin the provision of sustainable and meaningful ecosystem services by functioning as a holistic ecosystem.

Furthermore, the need for our urban areas and built environments to provide **significant** ecosystem services is one of the crucial challenges in Ecological Design. This is the fundamental core of ecomimicry. The aim is to minimise and eventually eliminate the parasitic dependence of the designed urban realm on ecosystem services provided by adjacent external environments, ecosystems from the wider rural hinterland of a nation or indeed ecosystems elsewhere in the world. This means working towards a macro scale intervention and aspiration in the integration and design of Nature-based infrastructure.

THE NEED FOR ECOSYSTEM BIODIVERSITY

The relationship between biodiversity and ecosystem function was discussed in Chapter 1. One example of the importance of biodiversity in the design of Nature-based infrastructure in the urban realm that is so often overlooked is the control of pests and diseases.

By thinking ecomimetically, we can design to ensure that a natural pest control strategy is achieved through the very biodiversity and intricate trophic levels within any created landscape system. Plagues of species that cause problems or irritation to us in urban areas are more often the result of impoverished or imbalanced ecosystems than because of the presence of natural habitats in the urban matrix.

AVOIDING AND ADDRESSING ECOSYSTEM FRAGMENTATION

From the point of view of the urban wildlife biologist, Nature-based infrastructure is green infrastructure, which is seen simply as the biotic network of habitat patches and the linkages used by wildlife to most safely commute between them. The linkages are often referred to variously as 'eco-passageways', 'habitat corridors' or '**wildlife corridors**'. Movement of species through the landscape is necessary for migratory species, gene flow, and population adaptation in response to changing environmental conditions (especially climate change) and natural disasters so that populations that go locally extinct (or approach extinction) can be replenished from surrounding areas.

Wildlife corridors are generally considered to facilitate the movement of both fauna and flora through a fragmented landscape in which much of the wider landscape matrix is inimical to their survival (e.g. because it provides little cover or food). In many cases, species will eventually disperse between habitat patches without the habitat corridors between them, crossing all manner of dangerous territory. Many will disperse between patches by air. However, for both land and aerially dispersing species, the degree of mortality is reduced where habitat corridors are present to provide safe passage between regions.

One good example is an endangered pangolin crossing a busy road bisecting two key rainforest remnants in Singapore. The animals will cross regardless of the inimical nature of the crossing, and many will die in the attempt. Creation of a vegetated bridge between two rainforest remnants has resulted in the reduction of such unnecessary premature mortality of this iconic species, and it has been shown that the pangolins are using the new human-made habitat as a safe conduit.

Other types of linkages formed by the fusion of technology and Nature include eco-underpasses, whereby ground-based fauna, often nocturnal, can move through the urban fabric with greater ease and safety.

Far more research is needed to prove the ultimate importance and utility of such corridors to the conservation status of many species.

For some specialised species that have particularly poor dispersal ability, spread from core patches of habitat may only occur in the presence of high-quality habitat linkages between patches (and then very slowly over many years).

Even when dependence on linkages or corridors for effective dispersal has been hard to demonstrate, such linkages have often been shown to serve, at the least, as supplementary habitats, supporting sink populations of individuals for which the main habitat areas may be larger patches. Such linkage areas can then act as reserves, buffering the larger habitat patches against environmental changes.

Clearly, such linkages also have many other functions in the urban realm, including being:

- Part of the hydrological infrastructure (see Chapter 6) and the physical part of the socio-political infrastructure used by people not only to get from place to place, but to do so in a way that provides the additional benefits of psychological well-being associated with biophilic responses.
- Key components in an open-air urban food production system (in addition to enclosed systems such as hydroponic farms).

Addressing the issue of fragmentation in Ecological Design of the urban realm, then, means very careful consideration of the two-dimensional connectivity (and three-dimensional: see next section) in the arrangement of biotic elements. This means the consideration of green network design and optimising this for ecosystem service provision in as many ways as possible (see next section). Careful consideration needs to be given to whether it is appropriate in any given ecological context to extend this network to connect to the wider countryside. In general, this is a good idea, but consideration needs to be given to the level of hazards that pertain in a given urban area at any one time and whether or not the time is ripe to encourage less-urban-adapted fauna into areas where their risk of mortality might increase. Similarly, the possibility of movement of introduced or alien species, often common in urban areas as they exist today, into sensitive rural habitats needs to be carefully assessed. The situation can change, and opportunities and options in this regard should be kept under constant review.

THE NEED FOR USE OF ALL URBAN DIMENSIONS

For many centuries, humankind has found elevated refuges from predators and foes that have also provided good opportunities for surveillance. These were often cliffs and mountains, frequently with cave systems in which our ancestors sheltered. Most of our food plants originated from scree slopes on mountain sides. These observations are the foundation of the so-called Urban Cliff Hypothesis. So perhaps it is not surprising that we build up and we build tall where we can in our cities. When it comes to provision of habitat and ecosystem services, then, it seems worse than negligent to neglect the vertical realm created by our urban structures. Facades, roofs and terraces all provide new opportunities for effective biointegration.

AUGMENTATION OF THE BUILT ENVIRONMENT TO PROVIDE ECOSYSTEM SERVICES

One of the most important aspects of Ecological Design is how we can emulate and replicate Nature's provision of ecosystem services. While this is referred to here as one of the most important attributes of ecosystems, it is necessary to note that as yet there is no demand by human society for humanity to have its societal systems, built environment and technologies provide ecosystem services. This may be because the provision of ecosystem services is not imminently and visually evident to humanity. However, by the time when this need becomes visually apparent, it may be that extensive loss of Nature's ability to provide ecosystem services will have been lost, which could have been averted much earlier.

Humanity needs to acknowledge with humility that Nature provides ecosystem services of such magnitude that so extensively cover all natural systems in the biosphere and are so complex in their production that they are impossible for any human technological system to be able to emulate and replicate in entirety. The solution to enhance the existing built environment's ability to provide ecosystem services is to physically augment the built environment and urban conurbations spatially with green infrastructures and by increasing biotic constituents and habitats interwoven spatially and systemically into the existing, mostly overwhelmingly inert, layouts of existing and new urban conurbations.

In applying ecomimesis to this endeavour, we need to change our built environment by making and remaking the existing features to have an equivalent biological structure to ecosystems. 'Biological structure' of ecosystems here refers to the existence of the biotic constituents together with abiotic constituents in an ecosystem that are integrated to form a whole system. The most sustainable, beneficial and multifunctional solution to enable humanity's built environment to provide ecosystem services is to elicit the help of Nature.

The approach is not to replace the services provided by ecosystems by technology alone, because of the magnitude and complexity of this undertaking. Rather than compete with Nature to provide ecosystem services, we need to elicit Nature to help provide ecosystem services in urban realms. The solution is to augment the built environment and technology to improve on what Nature has given us. This can be referred to as the '**ecological augmentation**' of the built environment.

Ecological augmentation (or 'ecoaugmentation') is where Nature is implanted into the built environment so that Nature merges with our machines and our built environment to co-exist harmoniously fused to function in an ideal biointegrated state indefinitely. The approach involves collaboration with Nature to augment our built realm and artefacts with natural resources to create hybrid multifunctional techno-biological constructed systems that have a complement of biotic and abiotic constituents that, as mentioned earlier, act together synergistically to become a whole system. Ecological augmentation, then, aims at the creation of **composite living systems** that are the seamless integration of Nature and technology, both on the ground plane and vertically with built structures (either new or existing). The resultant systems seek to be both biodiverse (in terms of species and processes appropriate to the locality) and capable of furnishing meaningful ecosystem services. To achieve this, great care needs to be taken in assessing the physical configuration, juxtapositions between elements and network design and patterns of the augmented systems.

This biointegration of biotic constituents has to go beyond just the physical installation of soils and vegetation into the urban environment and/or built systems. The biotic constituents need to be both physically and functionally **integrated** so that the combination functions holistically and displays **new emergent properties** as a result of the integration into a new 'constructed ecosystem'. This integration needs to make maximum use of the three-dimensional topography of our urban realm to create new opportunities for expression and maximisation of this integration.

The combination of Nature-based infrastructure by redesigning the urban realm needs to favour species adapted to the locality's climate and ecosystems in order to provide the underpinning and stabilising biodiversity within the resultant system. This means creating **meaningful habitats** for target species, which requires very careful consideration and application of specialist knowledge to a huge range of species, varieties, processes and interactions. The ecological designer needs to be aware of the complex processes and functions within the ecosystems (as summarised in Chapter 2).

Thinking of Nature-based infrastructure in this way means considering all of the life-cycle requirements of species, including food, shelter and places conducive to reproduction, which may entail the effective linking of several different habitats. But the additional element of systemic biointegration means fusing these habitat requirements physically and systemically with the mostly inorganic engineered urban constituents as a combined techno-natural system.

While, as stated earlier, geographic and physical scale is very important in provision of Nature-based infrastructure, simply increasing the areas of green space within our urban areas, for example by enlarging existent parks, green squares and gardens, is not enough. Many important species that once lived in the area will continue to shun the urban matrix.

To contribute to the survival of animal species that typically avoid urban areas but which are biogeographically appropriate to the locality is a particularly significant and worthy challenge for Ecological Design. To do this, we must truly understand habitat requirements and employ completely new design approaches.

One example of this is the potential use of **extra-normal stimuli** in design whereby a feature that typically acts as an attractant in natural or semi-natural ecosystems is exaggerated in an urban area such that a species actually prefers to use the human-made habitat.

Other examples of augmentation include the creation of artificial refuges for wildlife. Many such features that are poorly contrived are shunned by species, but others are hugely favoured and contribute very significantly to the local conservation status of species populations.

INFRASTRUCTURE DESIGN OPTIMISATION

The combined goals of maximising ecosystem service provision for target species of native flora and fauna in the designed urban ecosystem present many challenges. One of these is that the maximisation of one ecosystem service may be at the expense of another to some extent: e.g. the best design for hydrological management may not be the best for urban cooling. The next frontier in design of Nature-based infrastructure in combination with the other infrastructures is that of design optimisation. For this, multivariate computer models need to be developed to simultaneously assess the effects of design modifications on delivery of the widest range of ecosystem services and the greatest provision for target biodiversity. This challenge has already been taken up by institutions in various parts of the world and is at an exciting stage in the evolution of Ecological Design.

THE ROLE OF NATURE-BASED INFRASTRUCTURE IN ECOLOGICAL DESIGN

To sum up the core message, Nature-based infrastructure is Earth's life support system. It furnishes us, and all other species, with essential ecosystem services. It functions within thresholds of resilience at all geographical scales, from the macro to the micro.

In the planning and design of new urban developments, consideration of the retention, design and augmentation of Nature-based infrastructure should be first among the informing determinants of urban structure, not a bolt-on to what else is created. Where this has not occurred, retrofit of green and blue infrastructure is also possible. This entails the conversion of existing hard-engineered spaces and surfaces into biotic habitats, but this should be achieved by true biointegration that is both physical and systemic.

Before inserting or incorporating the green eco-infrastructure into existing cities and new built environments, it is crucial to study the ecological baseline conditions from site to bioregion. Key features of existing value need to be protected and the ecological integrity of key existing habitats supported. We need to preserve or create meaningful and viable habitats for local flora and fauna through our urban landscape designs, not only for the sake of Nature itself, but to underpin ecosystem service provision. Design needs to develop the site in a way that protects or restores these ecological services that link the land with its ecological history.

Ecological Design, then, really takes shape when we address the biointegration in space, time and system of Nature-based infrastructure with our technological, hydrological and socio-political infrastructures. From the resultant 'infrastructure matrix', we can examine and interrogate, and indeed model, the functioning of a combined quartrobrid constructed ecosystem with all its interactions. Nature-based infrastructure, then, truly includes the 'ecosystem-augmented built systems' that enable ecosystem services to be delivered optimally and at scale in the heart of the built environment.

NOTES

1 National Geographic. 'Biosphere.' *National Geographic Resource Library,*9 Oct. 2012, www.nationalgeographic.org/encyclopedia/biosphere/.

2 Liss, K. N., et al. 'Variability in Ecosystem Service Measurement: A Pollination Service Case Study.' *Frontiers in Ecology and the Environment,* vol. 11, no. 8, 26 June 2013, pp. 414–422. doi:10.1890/120189.

3 Andersson, E., et al. 'Reconnecting Cities to the Biosphere: Stewardship of Green Infrastructure and Urban Ecosystem Services.' *Ambio,* vol. 43, no. 4, 17 Apr. 2014, pp. 445–453. doi:10.1007/s13280-014–0506-y.

4 Hansen, R., and Pauleit, S. 'From Multifunctionality to Multiple Ecosystem Services? A Conceptual Framework for Multifunctionality in Green Infrastructure Planning for Urban Areas.' *Ambio,* vol. 43, no. 4, 17 Apr. 2014, pp. 516–529. doi:10.1007/s13280-014–0510-2.

5 Tzoulas, K., et al. 'Promoting Ecosystem and Human Health in Urban Areas Using Green Infrastructure: A Literature Review.' *Landscape and Urban Planning,* vol. 81, no. 3, 2007, pp. 167–178. doi:10.1016/j.landurbplan.2007.02.001.

6 Wirsen, C. O., et al. 'Chemosynthetic Microbial Activity at Mid-Atlantic Ridge Hydrothermal Vent Sites.' *Journal of Geophysical Research,* vol. 98, no. B6, 10 June 1993, p. 9693. doi:10.1029/92jb01556.

7 Berner, R. A. *The Phanerozoic Carbon Cycle: CO_2 & O_2*. Oxford University Press, 2004.

8 'Biogeochemical Cycles.' Environmental Literacy Council, 2018, https://enviroliteracy.org/air-climate-weather/biogeochemical-cycles/.

9 Cain, M. L., et al. *Ecology*. Sinauer Associates, Inc. Publishers, 2014 (Ch. 22).
10 Environmental Literacy Council. 'Sources & Sinks.' 2018, https://enviroliteracy.org/air-climate-weather/climate/sources-sinks/.
11 Lal, R. 'Carbon Sequestration.' *Philosophical Transactions of the Royal Society B: Biological Sciences*, vol. 363, no. 1492, 30 Aug. 2007, pp. 815–830. doi:10.1098/rstb.2007.2185.
12 Fierer, N. 'Embracing the Unknown: Disentangling the Complexities of the Soil Microbiome.' *Nature Reviews Microbiology*, vol. 15, no. 10, 21 Aug. 2017, pp. 579–590. doi:10.1038/nrmicro.2017.87.
13 Ernest, S. K., and Brown, J. H. 'Homeostasis and Compensation: The Role of Species and Resources in Ecosystem Stability.' Utah State University, https://digitalcommons.usu.edu/cgi/viewcontent.cgi?article=1344&context=biology_facpub.
14 Johnson, L. E. 'GIS and Remote Sensing Applications in Modern Water Resources Engineering.' *Modern Water Resources Engineering*. Humana Press, 2014, pp. 373–410.

FURTHER READING

Barker, G. 1997. *A Framework for the Future: Green Networks with Multiple Uses in and around Towns and Cities*. English Nature, Peterborough, UK. (Still often used as a standard reference in green space planning in the UK.)

Barton, H. 2017. *City of Wellbeing: A Radical Guide to Planning*. Routledge, London. (Recent review of the health benefits of green sustainable urbanism and how to achieve it.)

Collinge, S. K. 1996, Ecological Consequences of Habitat Fragmentation: Implications for Landscape Architecture and Planning. *Landscape and Urban Planning*, 36(1): 59–77. doi:10.1016/s0169-2046(96)00341–6.

Douglas, I., Goode, D., Houck, M. C. & Wang, R. 2011. *The Routledge Handbook of Urban Ecology*. Routledge, Abingdon, UK.

Dover, J. W. 2015. *Green Infrastructure: Incorporating Plants and Enhancing Biodiversity in Buildings and Urban Environments*. Earthscan, London.

Dunnet, N. & Kingsbury, N. 2008. *Planting Green Roofs and Living Walls*. Revised updated edition. Timber Press, Portland, OR. (Excellent practical guide.)

Earth Pledge. 2005. *Green Roofs: Ecological Design and Construction*. Schiffer Publishing, Atglen, PA.

EcoSchemes Ltd & Studio Engleback. 2003. *Green Roofs: Their Existing Status and Potential for Conserving Biodiversity in Urban Areas*. English Nature Research Report 498. English Nature, Peterborough, UK.

Larson, D., Matthes, U., Kelly, P. E., Lundholm, J. & Gerrath, J. 2004. *The Urban Cliff Revolution: Origins and Evolution of Human Habitats*. Fitzhenry & Whiteside, Markham, Canada.

Newton, J., Gedge, D., Early, P. & Wilson, S. 2007. *Building Greener: Guidance on the Use of Green Roofs, Green Walls and Complementary Features on Buildings*. Construction Industry Research Association, London. (Very useful practical guide to temperate systems.)

Oke, T. R. 1982. The Energetic Basis of the Urban Heat Island. *Quarterly Journal of the Royal Meteorological Society*, 108: 1–24.

O'Riordan, T. & Cameron, J., eds. 1994. *Interpreting the Precautionary Principle*. Earthscan, London.

Peterson, G. D., Allen, C. R. & Holling, C. S. 1998. *Ecological Resilience, Biodiversity, and Scale*. Nebraska Cooperative Fish & Wildlife Research Unit, Lincoln, NE. Available at http://digitalcommons.unl.edu/ncfwrustaff/4.

Stockholm Resilience Centre. 2015. The Nine Planetary Boundaries: What Is Resilience? Available at www.stockholmresilience.org/research/planetary-boundaries/planetary-boundaries/about-the-research/the-nine-planetary-boundaries.html.

Town and Country Planning Association. 2005. *Biodiversity by Design: A Guide for Sustainable Communities*. Town and Country Planning Association, London.

Town and Country Planning Association and the Wildlife Trusts. 2012. *Planning for a Healthy Environment: Good Practice Guidance for Green Infrastructure and Biodiversity*. Town and Country Planning Association, London.

Viljoen, A. 2005. *Continuous Productive Urban Landscapes: Designing Urban Agriculture for Sustainable Cities*. Elsevier, Oxford, UK.

SUN ENERGY
Food
Food
O2
CO2
Nutrient

6
Hydrological infrastructure

NATURE'S HYDROLOGY REGARDED AS AN INFRASTRUCTURE

The **hydrological infrastructure** is essentially all water-related components of Nature-based infrastructure as well as technological infrastructure. In the definitions of infrastructure set out in Chapter 2, it was emphasised that the categorisations are not mutually exclusive and are used largely to facilitate Ecological Design and to enable the endeavour to focus on a group of particular aspects of the fabric of our human-made world united by a common element. In the case of the hydrological infrastructure (or 'blue' infrastructure), this is water.[1]

Although water is, of course, the foundation of Nature-based infrastructure, it is treated as an 'infrastructure' in its own right because water, more than any other resource, is vital for the survival of all lifeforms on Earth. As organisms, we are composed largely of water and hence need to drink daily to stay alive. Currently, we rely almost totally on water for the processing and transport of sewage, we bathe in it, use it to purify, and all the other lifeforms that we depend on rely on water as well. In the days of easily accessible air travel, it is sometimes easy to forget how important water still is for transport. It remains vital for so many key urban areas across the world to be adjacent to the coast or a waterway or waterbody to aid transport of people and goods. Many of us in our day-to-day lives may also not think of how much water is used in industry and manufacturing. We are drawn to water very deeply psychologically, and a huge economic benefit derives from humanity's recreational use of water.

Water is also considered separately because, at the moment, it is not possible to produce potable water artificially at large scale without significant energy cost and environmental damage.

Accordingly, and as previously stated, the term 'hydrological (or blue) infrastructure' is used here to refer to that part of the natural infrastructure – the network of habitats, again both natural and human-made, and associated biogeochemical cycles that are based on aquatic ecosystems and the global hydrological cycles (and the term is normally confined to freshwater ecosystems) – combined with humanity's *built* hydrological systems. The built systems included here include water supply, drainage, and wastewater reticulation and treatment systems.[2,3,4]

WATER ON THE PLANET

Approximately 97.5% of the water on Earth is saltwater in our oceans, seas, bays, saline lakes and saline groundwater. We can use saline water, but only after we remove the salt, which takes considerable energy and infrastructure to achieve. When we do desalinate saltwater, we currently often discharge the waste hyper-concentrated brine into our coastal areas, often with very serious adverse effects on marine life.[5]

Of the 2.5% of Earth's water that is fresh, close to 70% is frozen in ice caps, glaciers, permanent snow, ground ice and permafrost. We can and do use some of this water too, but again, there are potential significant negative impacts. Melting ice is causing catastrophic sea level rise, impacting both coastal communities and ecosystems. Another major concern is thawing permafrost causing the release of methane, a far more potent greenhouse gas than carbon dioxide, catalysing effects of climate change.[6]

Of the remaining 30% of freshwater not locked away as ice, some 98.9% occurs as fresh groundwater, or in the soil. Over-abstraction from freshwater aquifers can have many adverse environmental effects. The process can create space that draws in seawater to replace the groundwater, contaminating the aquifer with non-potable water. The abstraction can affect surface wetland ecosystems very severely.

Of the just over 1% of freshwater unaccounted for, 90% occurs in the Earth's rivers, lakes and swamplands, and the remaining 10% is held in the atmosphere. We can use all such water, but again with environmental consequences for vital freshwater ecosystems on which we rely for many other goods and services, and getting water from the atmosphere is possible, but very inefficient in most circumstances.

Water locked up in biological organisms accounts for only about 0.01% of global freshwater.

Overall, it has been estimated that currently less than 0.001% of global water is readily available for human use. So it really is almost a case of, as in the poet Samuel Taylor Coleridge's *Rime of the Ancient Mariner*: 'Water, water everywhere, / Nor any drop to drink.'

In many countries, for all of the reasons listed above, there is a tremendous scarcity of utilisable safe, freshwater. Water resource management is made particularly difficult by the fact that exploited sources of freshwater often cross national boundaries. Wars have already occurred in parts of the world over remaining freshwater resources. Often, it is very difficult to secure accurate information on where freshwater occurs in the ground, how much there is and what effects its abstraction might cause. When treated as a commodity internationally, water creates problems of valuation in traditional economic terms.[7]

The paramount importance of addressing the Planet's water resources and the hydrological infrastructure by which they can be managed sustainably is clear. The hydrology of the built environment and the hydrology of the natural environment are totally intertwined by the **water cycle** (see Glossary). Considering them together holistically is vitally important so that we 'close the local water cycle' at the interface between built and unbuilt (natural or semi-natural environments). This means ensuring that precious freshwater is kept recycling healthily and sustainably in benign local systems rather than being constantly abused and discharged offsite as waste.

WATER RESOURCE MANAGEMENT

Water resource management in the not too distant past was solely concerned with ensuring continuity of supply of potable and other water suitable for human needs. Where pollution of the water environment occurred, concern was really limited to the security of supply rather than ecosystemic harm. In more recent decades, the whole concept of integrated water resource management has emerged, where the key metric of how sustainably we are using our precious and limited freshwater resources is the conservation status and health of aquatic ecosystems.

The fact sadly remains, however, that in the majority of human-built environments worldwide, management of water is limited to the creation of huge water stores, abstraction from water bodies, derogation of surface wetlands due to groundwater abstraction, and incessant and often incredibly damaging pollution.

We are also profligate in our use of water. A family household in the USA might use almost 2000 litres of water per day and toilet flushing alone accounts nearly for 30% of that.

Water management in an ecomimetic and ecocentric world needs to ensure that our natural wetlands, oceans and other water sources continue to survive in healthy condition such that they can continue to show full ecosystem functionality and provide the full suite of characteristic ecosystem services. Water demand henceforth needs to consider all of Nature, not just human beings.

Ecological Design as ecomimesis, therefore, in emulating ecosystems hydrology, needs to show much greater efficiency in use of water, and achieve a positive water balance where the water taken from the environment needs to be returned to it in as pure a form as, or an even purer form than, when it was abstracted.

KEY GOAL: MINIMISATION OF POTABLE WATER USE

Use of potable water is of particular concern given the high carbon footprint so often associated with its production (e.g. from wastewater) and transport from source to use. The traditional techno-centric solution is to focus exclusively on engineered ways to develop new supplies of water, such as super-sized dams that remove many of the ecosystem services provided by our most important rivers and huge aqueducts, and pipelines that deliver water over very considerable distances. Key methods by which we can do this include:

- Water economy mechanisms in the home, such as smart flow meters, aerator taps and reduced flush toilets.
- No-dig agriculture and permaculture to ensure maintenance of good water-retaining soil structure to reduce demand.
- Smart irrigation systems in agriculture, including cybernetically controlled surface, and ideally sub-surface, irrigation systems, timed to deliver water exactly when and where it is needed with minimum waste.
- Providing alternative sources of water for human uses that do not absolutely require potable water (e.g. toilet flushing or irrigation of gardens).

KEY ECOLOGICAL DESIGN GOAL: A PROPERLY FUNCTIONING WATER CYCLE AND WATER SUPPLY

Fluxes of water

Water moves in mysterious ways, transforming to vapour and reforming as water, freezing into ice, flowing through our watercourses and travelling airborne through clouds. It also moves through plants from roots to body to leaves and stems, where it is excreted as a by-product of photosynthesis in a process called transpiration.

Water vapour in the air condenses as droplets, often around dust particles, and then, whether near or far from the site of cloud formation, eventually precipitates as rain, sleet, hail or snow. Some escapes from the soil surface (infiltration) through the soil profile or pores and fissures in permeable rock (percolation), and ultimately to groundwater. The rest either evaporates or travels via watercourses to end up in our lakes, coasts, seas and oceans.

All of these different compartments of water are in continual flux, one with the other. The patterns of movement and the condition of key physical 'vessels' for this water, such as rivers and lakes, can be radically altered by human activity.

The peaked 'urban hydrograph'

Conventional design of the water infrastructure in our urban areas has led to increasingly large areas of impermeable surfaces comprising roads, parking lots, pavements and roofs. These prevent water infiltration and percolation of rainwater. Instead, rainwater runs off these surfaces. If the drainage network cannot accommodate this runoff, it can lead to local pluvial flooding of urban areas, and this often occurs at a time when the risk of flooding from rivers bursting their banks is also high.

A particular problem comes when the sewer system is used as a combined sewage overflow system with stormwater. Normally, this may not be a problem (though it is a huge waste and degradation of a resource), but in the case of heavy rains and storms, the whole system will overflow, with combined flows of rainwater and human effluent running off untreated into surrounding waterways. This takes raw untreated sewage, albeit somewhat diluted, directly into some of our most sensitive ecosystems, which also causes a public health concern for recreational use of the water.[8]

Not only is the pollution load a problem (see later in this chapter), but the volume and speed of flow of water entering the watercourses and waterbodies does further damage, eroding aquatic and riparian habitats, and hence creating a big ecosystemic shock.

The problems caused by this '**peaked hydrograph**' (see Glossary) of water runoff from urban areas will worsen, as future intensity of rainfall and frequency of storms are generally predicted to increase substantially.

Soil water

This diversion of rainwater also means that less rainwater can infiltrate and percolate through the soil. This in turn means less recharge of groundwater, which in most urban areas around the world is already being heavily abstracted. One result of this is the desiccation of the roots of urban trees and other plants and adverse effects on plant health.

In Ecological Design of our urban areas, we need to ensure that there is sufficient soil moisture for healthy plant growth. Other factors placing high stress on plants in the urban environment include relatively high temperatures due to the urban heat island effect (see Chapter 5). If plants have enough water to maintain turgor pressure (cell pressure in all parts of a plant), they can better cool the urban environment, not only by shading with healthy leaf canopies, but also by more efficient evapotranspiration.[9]

Trees can grow tall enough to 'roughen' the urban air flow, allowing for mixing and convection of air masses so that hot air can escape the urban matrix.

It is a common misconception that what most trees need is good soil to grow. Nutrients certainly help, but trees like birch and rowan can even colonise the skeletal soils of glacial moraines. What they need most is a good water supply and reasonable temperature regime.

Groundwater resources and recharge

In places where groundwater and soil moisture are related to one another directly, maintaining groundwater helps to maintain soil moisture for the growth of vegetation, with all the associated benefits that brings to our urban realm (Chapter 5). Other benefits include the potential effects on more distant ecosystems that may be groundwater-fed.

As mentioned earlier, the situation in most urban areas on Earth is one of significant groundwater abstraction for direct and indirect human use, very little of it returning to the same aquifer. In some cases, the aquifer abstracted is very deep, sometimes many hundreds of metres below the surface of the ground. Some of these water resources are of ancient water that fell as rain main millennia previously. Some are no longer naturally recharged at all. Others are, but the processes that recharge them, often very great distances from their point of use, may themselves be compromised. Clearly, these resources are finite in any practical sense, and need to be cared for, and where possible, restored as would occur in any natural ecosystem (see later in this chapter).

Technological solutions can be devised to take the place of Nature. These are generally referred to as 'artificial groundwater recharge of aquifer injection'. The use of such techniques, however, is a good example of a contrived solution to a problem we should not be creating in the first place. Moreover, the process of natural percolation down to groundwater is in general far less risky environmentally than direct injection of groundwater by boreholes and pumps.

KEY GOAL: ELIMINATING POLLUTION OF GROUNDWATER AND AQUATIC ECOSYSTEMS

Clearly, the scarcity of readily utilisable valuable freshwater means that any additional factor that makes it less readily usable should be avoided.

As emphasised in Chapter 5, we have traditionally used our rivers, lakes and oceans as waste disposal systems. The effects of nutrient enrichment have had devastating effects on our freshwater and marine ecosystems (excess phosphorus and nitrogen being the primary culprits). The excessive input of nutrients comes mainly from agricultural runoff containing high-nutrient fertilisers used excessively, as well as runoff from poorly treated wastewater systems containing traces of personal and home care products washed down the drain. The presence of excess nutrients leads to accelerated growth of algae, commonly referred to as a 'harmful algal bloom', which then consumes dissolved oxygen in the water, creating hypoxic conditions and blocking sunlight. Without oxygen or light, much of the remaining aquatic life will suffer or die. Such conditions also favour many pathogens and vectors of disease, including mosquitoes. This process is called cultural eutrophication.[10]

Our wastes discarded to both surface water and land have contaminated groundwater extensively. Raised nitrate levels in groundwater as a result of agricultural runoff have been linked to brain damage in human infants when their key water source is abstracted groundwater.

Technological solutions exist for many forms of aquatic pollution, and more are being developed as part of a vast global research effort. But the simple question the ecological designer needs to ask is why create a problem that then requires a costly solution or which will store up other difficulties for the future?

Waste processing systems can be very elaborate. An entire global industry of huge scale has arisen to address the technical treatment of human waste, including industrial and agricultural effluent and urban **greywater** and **blackwater** (see Glossary). Are there not better approaches?

ECOMIMETIC APPROACH OF CATCHMENT/ WATERSHED MANAGEMENT

In previous times for humanity, the means of addressing the combined flood risks of fluvial, pluvial and coastal (saline) flooding has been to build higher flood defences, canalise water courses to get water 'away' as rapidly as possible and to keep land generally as dry as possible. In recent decades, however, the awareness has risen steadily (though not yet universally accepted by all elements of society) that a holistic ecological approach is instead needed, working at the catchment/**watershed** scale (see Glossary). This awareness has been enhanced by the apparent recent increase in frequency of severe weather events and flash floods that were not fully predicted by those responsible for ensuring public safety from flooding.

The problem in so many instances is one of unfavourable land use change within a given catchment, especially in its areas of highest topographical elevation. Where tree cover has been removed, for example to raise grazing animals (which also compact the surface of the soil), the water runoff is greater in volume and more rapid (see later in this chapter), and this can lead to floods in the lowlands, especially in valleys along watercourses where so many human settlements have been historically established.

The topic of catchment management is huge, and the literature and research burgeoning. But the fact remains that we continue, by and large, worldwide to try to solve problems in one place that we have created in another by not taking a holistic, ecomimetic approach.

ECOMIMETIC APPROACH OF SUSTAINABLE DRAINAGE SYSTEMS

Ecomimetic design as applied to the issues of runoff management has the goal of restoring local hydrological processes as near as possible to natural or semi-natural conditions (often referred to as 'greenfield' conditions).

In response to these issues in our urban and suburban areas, in relatively recent times designed water management systems that to varying extents mimic natural ecosystems have arisen. The name given to these systems is sustainable drainage systems (SUDS). The aim of SUDS generally is to slow down the rate of runoff, flatten the peak of the urban hydrograph, and filter and treat the runoff through biotic processes.

Pollutants in runoff vary, but may include metals, oils, soil particles, excess nutrients and microorganisms, including viruses. Rainwater can contain many of these pollutants before it even hits the ground because it picks up pollutants in the atmosphere.

Many SUDS are ecomimetic systems that are combinations of some physical excavation or supporting structure and water-tolerant vegetation. SUDS have a variety of mechanisms to both purify the water passing through them and as attenuate it. In general, it is the substrate in which the vegetation is growing that does most of the water treatment. Bacteria and other organisms in the substrate utilise or break down nutrients, metals and other chemical contaminants in the water. The substrates are kept functional, porous and often oxygenated by the plants growing in them. The plants themselves can also bio-accumulate pollutants to varying extents.

The contaminants are processed and filtered out as water moves through the so-called 'treatment chain' of interconnected SUDS. Efficiencies of removal vary, but can be very high for contaminants such as suspended solids and heavy metals, and quite good for plant nutrients like nitrogen and phosphorus, depending on the design of the SUDS and the overall 'retention time' – that is, how long the water stays in the overall SUDS system.

Plants for SUDS may be specially selected for their bioaccumulation abilities. However, it has been shown that a good variety of native wetland plants over a long treatment chain in sequence can produce excellent levels of runoff treatment. This has the advantage of being best for native biodiversity, and often visual amenity too. Use of non-native wetland plants in SUDS is generally not recommended, as spread of plants between wetlands and along watercourses is especially rapid, and if any alien plant becomes invasive, it can have devastating effects on native wetland ecosystems and their flora and fauna.

While 'created wetlands' may not have the biodiversity value of natural wetlands, largely because they have to sustain the continued environmental shocks of the runoff, if carefully designed, they can become vital parts of the designed urban system, and indeed very high-quality Nature reserves.

EXAMPLES OF SUSTAINABLE DRAINAGE SYSTEMS

The list of systems that are hybrids of technology and Nature that are used as part of the SUDS 'armoury' has increased steadily over the years, as has the nomenclature referring to them. A few examples will suffice to demonstrate this.

Existing green spaces

It may seem rather obvious that existing green spaces are part of SUDS. However, the SUDS functionalities of existing urban green spaces can also be increased by directing stormwater to them, which may cause flooding, but at a time when there is no concern over loss of amenity.

Constructed wetlands

There is a bewildering variety of these. They range from modifications to existing green spaces by creating bowls of depression for the collection of water to highly contrived systems with artificial substrates and water control pipework. These range from so-called 'vegetated French drains', 'bioswales', 'attenuation basins' and 'detention basins' to fully fledged larger-scale treatment wetlands. The type of biological SUDS installed in a given location depends on a variety of factors, including position in the so-called 'treatment chain' and the availability of space.

The aim is to start attenuating and treating the rainwater from the minute it hits the highest parts of the urban fabric. Thereafter, either attenuation or treatment may be the most important function to achieve at any given location. These contrived habitats can also be multifunctional, providing attractive visual and sometimes physical amenity in the case of features that are only filled with water during storms, wildlife habitat and local cooling of the urban fabric.

Green roofs and facades

Green roofs and facades have been increasing in popularity for some decades, but only recently in some areas have they been accepted as fully functional components of SUDS systems. Green walls have had a more chequered history, but their potential as significant SUDS, especially in places where there are greater limitations on groundspace, is now being seriously explored. One excellent example of this is the Rubens Hotel near Victoria in London, which can attenuate and filter very significant quantities of roof runoff, while growing a wonderful array of native and naturalised plants that provide great opportunities for pollinators.

Hard SUDS

These sorts of SUDS are basically structures which provide storage in interstitial spaces and cells of their hard structures (e.g. porous pavements and roads, or supercells) without any surface vegetation forming part of the system. This might not seem very ecomimetic, but there is biological treatment of the water through the action of biofilms of microbiota with the storage fabric/cells.

ECOMIMETIC APPROACH OF RAINWATER HARVESTING

Once water is drained into rivers, lakes or into the sea, it becomes very costly to recover and process into potable water. However, ensuring passage of sufficient water to natural waterbodies and watercourses, and indeed to the oceans, is vital to maintain the ecological integrity of those systems, so it is incorrect to consider such water as 'wasted'.

As long as this requirement for maintaining the integrity of aquatic ecosystems is met, there is scope for longer-term storage of water in urban catchments. This is known as 'rainwater harvesting'. Indeed, carefully undertaken, this can help balance out periods of storm and drought not only in the urban realm, but in some offsite ecosystems too.

When SUDS and rainwater harvesting systems are combined, often controlled by cybernetic monitoring systems, the approach is referred to as 'water-sensitive design'.

ECOMIMETIC APPROACH OF UPSCALING URBAN SUDS TO 'SPONGE URBANISM'

In recent years, a new term has arisen in China to refer to the large-scale adoption of SUDS systems and water-sensitive design in urban areas. This is the concept of 'sponge urbanism' or the 'sponge city'. All the functionalities of SUDS as just described are writ large and systematically, creating carefully contrived and designed integrated networks across the city.

By designing SUDS approaches as the town or city scale, it is easier to ensure the provision of significant multifunctionality and to consider how to optimise a variety of ecosystem services over a large area. This in turn can help to facilitate sustainable water planning over entire metropolitan regions and their hinterlands. It can also start to lead to meaningful budgetary adjustments in terms of the provision and maintenance of hard drainage infrastructure.

The value of working at scale is that any ecosystem service provision, including flood risk reduction, real urban heat island reduction as opposed to just local cooling, replenishment of clean groundwater, and ecopsychological benefits from biophilic responses (which are particularly associated with the water environment: see Chapter 2), are greatly increased by the extension of the 'biophilic diet'. This means the greater the exposure to the features that produce the biophilic response in space and time, the greater the measured benefits, and a city-scale SUDS network will certainly help in that regard. Benefits to native biodiversity are also greater the higher the scale of intervention, for the reasons explained in Chapter 2.

ECOMIMETIC APPROACH OF INCORPORATING GREYWATER RECYCLING

We pollute the water we use each day domestically to varying extents. When the pollution does not entail significant quantities of human excreta, then the resultant polluted water is generally referred to as 'greywater'. This is the water that is usually discarded to sewers after we have washed ourselves, our clothes and our cooking utensils. Greywater, although not as dangerous as blackwater (see later in this chapter), does carry a bacterial load and often considerable concentrations of plant nutrients (as detergents are most frequently based on phosphates).

The recycling of greywater is totally feasible. For example, in California, where there are significant problems with drought, it is now mandatory to divert sink and bath water to toilet cisterns rather than to flush toilets with precious potable water.

Systems have been developed for the modest urban scale (low-rise blocks of flats) to divert greywater to roofspace, filter it through rooftop SUDS, apply final ultraviolet treatment as a precaution to kill residual pathogens and then add a vegetable dye so that it cannot be easily confused with potable water before it is reused for toilet flushing or irrigation of gardens.

Quite often today, arguments are made that the capital cost of a third water system in addition to potable and sewage systems is uneconomic in terms of capital cost. Quite often, changes to address this sort of impasse only need to be kick-started through a combination of socio-political and economic mechanisms (see Chapter 8). The mass scale of production of the systems then reduces costs, and before long, the technology becomes a standard part of build costs.

ECOMIMETIC APPROACH OF RECYCLING OR ELIMINATING BLACKWATER

At present, we deal with human excreta by diluting them, generally with potable water (though see earlier in this chapter), and turning them into a waste product: sewage. From a practical point of view, especially at high human densities, it is clear why water is used as the conveyance medium as it is already the main constituent of the excreta. Looked at in more theoretical and holistic terms, this is an impossible thing to do. It is like taking two resources and mixing them together to create a problem waste product.

Human urine does not contain a high level of pathogens as a rule, and relatively simple biotic systems can treat it safely. Human faeces, with careful composting for sufficient time above 60 °C, can become a relatively pathogen-free fertiliser. Potable water is, as discussed earlier, a precious commodity won often at considerable cost.

At small scale in 'eco developments' such as in the so-called 'Hockerton' housing development in the UK, separator toilets are installed. These separate urine from faeces into different routes. The urine is treated in a fairly natural wetland very safely. The faeces are composted locally and on a small scale until safe, and then used as fertiliser for homegrown crops.

Human waste can also be composted at larger scale using anaerobic digestion tanks, which produce methane gas that can be combusted as fuel. This may entail mixing human excreta with, for example, domestic green waste and the waste associated with stock animal husbandry. The most sustainable benefits of this obtain when the combustion takes place in **combined heat and power plant** (see Glossary), so that the greatest amount of resultant energy is reutilised.

Doing this at scale for large towns and cities, however, is rather more problematic. The sewage in our cities enters a sewer network and is generally taken to sewage treatment plants that use a whole slew of physical, chemical and biological processes to treat it to a point at which the original organic matter and water are semi-separated again.

The aqueous effluent can then be relatively safely released into the environment (typically rivers). The safety of this is only relative, as one must not overlook potential ecosystem damage. At present, in most cases there are no or very high limits to the phosphorus loads in the discharges from such treatment works, and as was explained earlier, this is experimental for natural and semi-natural wetlands in terms of eutrophication risk. Phosphate stripping plants can be included, at notable cost – another example of an expensive 'technofix' to a problem that we should not perhaps be creating in the first place.

The other main product is a residual organic sludge. This can be used as fertiliser. However, there are problems here too. Due to the mixing of urban runoff and human sewage in most of our urban areas, the sewage contains not only human excreta, but contaminants such as various types of metals that then end up in the sludge. Moreover, due to the ignorant and sometimes illegal actions of many in society, the sewage can become contaminated by other dangerous compounds such as solvents and cancer-causing agents discarded down toilets and drains. This means that using the sludge to fertilise crops for consumption may not be safe. Accordingly, sludge is often disposed of as a fertiliser for forestry.

All the above problems are made worse where industrial waste products are also discharged to the sewer system without adequate pre-treatment. This is a very severe problem in some parts of the world. All in all, the current loop of blackwater waste creation and treatment, especially when many developing countries have no treatment systems in place, is an inefficient and ineffective approach with harmful repercussions.

Are there more ecomimetic approaches? It would seem that if one is going to create waste, treating it as close as possible to its source of production might be preferable. Moreover, rather than having partly clean water as the end product, what about potable water? In Belgrade, a large internal greenhouse system with multiple artificial biomes and carefully contrived cultures of bacteria that treat the **population equivalent** (see Glossary) of the sewage from 50,000 people. There is no adverse odour, and the installation is a visitor attraction. Products of the system include drinking water and a variety of other utilisable products at a relatively high degree of purity. This demonstrates a step towards a more ecomimetic future.

TECHNOLOGY IN THE MANAGEMENT OF ECOMIMETIC HYDROLOGICAL INFRASTRUCTURE

It should be clear by this point that managing our hydrological infrastructure involves the careful control and integration of a significant number of different aqueous flows, processes and by-products. The systems that do this are already a step along the way to true integration of engineering technology and biology, though they could go much further in this regard.

One of the key areas in which improvements can be made is in terms of multifunctionality. For example, sewage also contains waste heat generated in the human bodies that produce it and the dwellings in which it is generally produced. This heat is generally wasted, but can be recovered and reutilised with appropriate engineering. Integrating and optimising all of the various elements of the hydrological infrastructure so that processes are as ecomimetic and optimised for as many functionalities as possible cries out for the integration of multivariate cybernetic computerised control systems. The future marriage of the cybernetic and the technological in ecomimesis will be discussed in Chapter 7.

ENCOURAGING POSITIVE ANTHROPOCENTRIC PARTICIPATION

One of the risk factors in the sustainable and ecomimetic operation of any hydrological infrastructure is *Homo sapiens*. As stated above, our actions are not always optimal. As with all approaches in sustainable design, the goal must be to make the right thing the simplest and most attractive thing to do. This can be achieved through a combination of incentives (e.g. tax and insurance breaks) and punishments (e.g. tariffs). But it is also vital to continue to educate and raise awareness. One way to do this is to make the flows and processes within hydrological infrastructural systems apparent and patent, for example through real-time graphic displays in homes or public places. This may empower and galvanise citizens to embrace complex hydrological ecomimesis as an ultramodern and exciting advancement for humanity, helping to combat the frequently held public view that ecological approaches are somehow reactionary and anti-progressive.

SUMMARY OF HYDROLOGICAL INFRASTRUCTURE FACTORS

Clearly, the aims of ecomimetic hydrological infrastructure design is to ensure that the systems are multifunctional, run on sustainable energy (ideally provided directly by the sun) and have all the resilience and homeostatic properties of natural ecosystems (see Chapter 2). It is a comprehensive approach that includes conservation, efficiency and sensible frugality by humanity. As emphasised in Chapter 4, a redesigned ecomimetic hydrological infrastructure needs to be biointegrated with the other infrastructures – Nature-based infrastructure, socio-political infrastructure and technological infrastructure – to create a truly sustainable human-made world that may save our future and that of other species.

NOTES

1 Edwards, P. E. T., et al. 'Investing in Nature: Restoring Coastal Habitat Blue Infrastructure and Green Job Creation.' *Marine Policy*, vol. 38, no. 12, June 2012, pp. 65–71. doi:10.1016/j.marpol.2012.05.020.

2 Karamouz, M. *Hydrology and Hydroclimatology: Principles and Applications*. CRC Press, 2013.

3 Margulis, S. *Introduction to Hydrology*, 4th ed., 2017, https://margulis-group.github.io/teaching/.

4 Dang, T. D., et al. 'Hydrological Alterations from Water Infrastructure Development in the Mekong Floodplains.' *Hydrological Processes*, vol. 30, no. 21, 29 Apr. 2016, pp. 3824–3838. doi:10.1002/hyp.10894.

5 Elimelech, M. 'Overview on Nanotechnology: The Future of Seawater Desalination.' *International Journal of Science and Research*, vol. 5, no. 2, 2016, pp. 399–401. doi:10.21275/v5i2.nov161131.

6 O'Connor, F. M., et al. 'Possible Role of Wetlands, Permafrost, and Methane Hydrates in the Methane Cycle under Future Climate Change: A Review.' *Reviews of Geophysics*, vol. 48, no. 4, 2010, doi:10.1029/2010rg000326.

7 Reisner, M. *Cadillac Desert: The American West and Its Disappearing Water*. Penguin, 1993, pp. 1–47.

8 Eregno, F. E., et al. 'Quantitative Microbial Risk Assessment Combined with Hydrodynamic Modelling to Estimate the Public Health Risk Associated with Bathing after Rainfall Events.' *Science of the Total Environment*, vol. 548–549, 2016, pp. 270–279. doi:10.1016/j.scitotenv.2016.01.034.

9 Yurtoğlu, N. 'Plant Water Relationships and Evapotranspiration.' *History Studies International Journal of History*, vol. 10, no. 7, 2018, pp. 241–264. doi:10.9737/hist.2018.658.

10 Smith, V. H. 'Cultural Eutrophication of Inland, Estuarine, and Coastal Waters.' *Successes, Limitations, and Frontiers in Ecosystem Science*, 1998, pp. 7–49. doi:10.1007/978-1-4612-1724-4_2.

FURTHER READING

Cantor, S. L. 2008. *Green Roofs in Sustainable Landscape Design.* W.W. Norton, New York.

Gunawardena, K. R., Wells M. J. & Kershaw, T. 2017. Utilising Green and Bluespace to Mitigate Urban Heat Island Intensity. *Science of the Total Environment*, 584–585: 1040–1055.

Wells, M. J. & Grant, G. 2006. Biodiverse Vegetated Architecture Worldwide: Status, Research and Advances. In: *Proceedings of the 22nd Conference of the Institute of Ecology and Environmental Management: Sustainable New Housing and Major Developments – Rising to the Ecological Challenges.* Institute of Ecology and Environmental Management, Winchester, UK. Available at www.livingroofs.org/images/stories/pdfs/WellsGrant%20_final.pdf. (Historical overview of green roof design.)

TUCK
respiration
photosynthesis
organic carbon
dead organisms & wastes
fossils & fossil fuels
auto & factory emissions

7

Technological infrastructure

WHAT IS TECHNOLOGICAL INFRASTRUCTURE?

The term '**technological infrastructure**' refers to all of human society's synthetic and semi-synthetic artefacts, built structures, infrastructures and built systems that humans make, construct and manufacture from raw materials found on Earth. These raw materials include both inorganic raw materials extracted from the Earth (e.g. metals and minerals) and organic raw materials that grow and are later harvested (killed and gathered) for use in their non-living state (e.g. timber).

External and internal infrastructures in the cities and urban conurbations

Although society's synthetic and semi-synthetic artefacts are often created for commercial, domestic and other uses, the largest of all human artefacts are cities and urban conurbations. The buildings and miscellaneous structures, such as bridges, ports, pylons and all the parts of a typical urban conurbation, make up these megastructures.

Urban utilities are the connected technological *infrastructures* that can be sub-categorised as those that are **external** to enclosures and buildings. Utilities are either hidden in the ground, raised in the air by pylons, or exist as **internal** utilities, such as within buildings and usually hidden within the walls and ducts in the built structures above ground. The urban technological infrastructures are as follows:

External technological infrastructures

These include all networks of engineered utilities that are installed largely outside buildings. Examples include:

- Road and routing systems, both above ground and underground, such as rail lines, and the vehicles that utilise them (cars, trains, buses, trucks, etc.).

- Miscellaneous structures that are part of the urban realm such as bridges and tunnels.

Underground transport systems;
Docking areas, parts, airports, wharfs;
Underground and undersea cables;
Overhead power and telecommunication pylons and lines;
Vacuum waste collections systems.

- Sewerage lines and treatment systems.
- Water reticulation systems.
- Energy generation systems and electricity power lines.
- Telecommunication and IT systems.
- Urban gas systems.
- Other urban utilities.

For the sake of completeness, the conceptually mapped and regularly used routes for movement of people and goods by sea, air or space must be included, along with the vessels, aircraft and spacecraft that follow them. It may seem strange in a way to think of these amorphous media as 'infrastructure'. But in fact, we use many routes so often that our use leaves chemical and other 'footprints' such as the vapour trails of aircraft that have a significant effect on weather and climate.

Internal infrastructures in enclosures

These are the engineering systems that are installed within buildings and built structures to enable humans and human-made machinery to function effectively and efficiently within the enclosures – that is to say, within any architecturally enclosed systems, building envelopes or other structures. These are often buildings massed laterally or vertically as super-enclosures. Within them, human society takes shelter from the elements to live, reproduce, work, manufacture and recreate. These are the mechanical and electrical systems/services, of which there are a bewildering array in the modern building. They include:

- Gas networks.
- Electrical networks.
- Water supply system for potable water.
- Sewerage and wastewater systems.
- Machinery and appliances for cooking and lighting, heating and cooling.
- Fire-fighting systems.
- Mechanical movement systems such as elevators and escalators.
- IT and telecommunications systems, appliances and cables.
- Environmental heating and cooling systems.

The enclosed infrastructures and enclosures enable humanity to carry out all of its domestic, commercial, industrial, productive, economic, political, cultural and various societal functions within environmental conditions that are to varying extents weatherproof and climate-regulated to enhance human comfort, health and productivity.

Most of human society gives little thought to these internal networks of infrastructure, not least because they are generally hidden within the walls or roofs of enclosures rather than being clearly visible (although there are exceptions).

Many of the components of these infrastructures are essential for our survival and for the efficient functioning of modern society. However, there may be some which may not be so essential, and some are even beginning to harm us (such as our addiction to many products of information technology: see Chapter 8).

These infrastructural systems, both externally and internally, can contribute to negative human impact on Earth's systems and cycles, and need to be redesigned systemically, and in many cases retrofitted, to enable society to be ecologically effective (see Chapter 2). These systems need to be designed and made to be ecologically benign to Nature and be based on ecomimicry of ecosystems attributes. For example, the energy systems need to be powered by renewable sources in a similar way that the most ecosystems generally rely on the sun as the primary renewable source of energy. Water sources need to be local and used in a replenishable way. Wastes need to be recycled.

These urban technological systems and their hardware need to have low embodied energy in their content and low energy in their operations, achieving net zero energy if possible and be reusable, recyclable and integrated back into nature.

The challenge in Ecological Design is to design and make these infrastructures and structures based on ecocentric and ecomimetic principles.

THE MULTITUDE OF ARTEFACTS IN HUMAN SOCIETY

Included in the technological infrastructure and built environment are the multitude of artefacts, items, machinery and equipment that humans make as part of commercial operations, industrial production and domestic use. These include, for instance, computers, white goods, clothing, toys, mobile phones and things that are made for use for human daily lives and well-being. While there are considerable numbers of artefacts already in existence in human society whose disposal and recycling need to be addressed and remade, the making of all new artefacts needs to be based on ecomimicry, so that they are made to be reusable, recyclable and to be easily returned back into the environment without doing harm to Nature.

Do other animals make artefacts? There are indeed other species of animal that manufacture artefacts, but the degree of processing of natural raw ingredients that occurs in the manufacturing process is limited and biodegradable.[1] This means that that the resultant artefacts can be generally be benignly absorbed back into the biospheric ecosystems through decay, weathering and other processes.[2,3,4]

For example, weaver birds and bower birds produce complex nest structures that are true miniature architectural marvels, but which decay relatively quickly.[5]

Dams created by trees felled by beavers in California have been found to date back more than 1000 years because beavers constantly renew the structures in the same locations. Accumulated vegetation begins to decompose, creating anaerobic environments (with no oxygen), which promotes bacterial life that produces methane. The scale of methane production ultimately caused by beaver dams is sufficient to contribute significantly to the global production of greenhouse gases.[6]

Termite mounds can also remain intact over impressive timescales. The mounds formed by these amazing ecological engineers are the combination of soil, sand, clay, wood and other natural materials bound together by their sticky saliva. The many mounds built by the now extinct termite *Syntermes dirus* in the Caatinga deserts of north-eastern Brazil form strange lunar landscapes of well-spaced bumps that have endured for nearly 4000 years. Below ground, the structures are even more extensive.[7]

Coral polyps (communal animals that look similar to inter-linked sea anemones) create vast structures of coral reefs formed from (among other things) biologically produced limestone. Coral reefs include some of the largest animal-made structures on Earth: the Great Barrier Reef on the east coast of Australia is visible from outer space.[8,9]

Our Stone Age ancestors made and shaped tools from rock which have endured for tens of thousands of years. However, at one point *Homo sapiens* breached a threshold in our growth as we discovered how to create and manipulate fire. From that point, our manufacturing capability exponentially developed, and we started to create chemicals and materials that were of rare occurrence on the Earth, certainly in the biosphere; many had not occurred naturally before in the biosphere.

As our technological abilities have grown (as will be discussed further later in this chapter), so has our ability to synthesise ever more extraordinary non-natural chemicals with absolutely enormous disruptive power in natural ecosystems. Fortunately, no other animal does this as extensively as humans.

PROBLEMS WITH OUR LARGEST ARTEFACT – THE CITY

In Chapter 5, the very limited extent of Nature-based infrastructure in so many urban areas across the Planet was discussed at length. The Nature-based infrastructure – the combination of original, highly modified and newly created biotic systems supporting vegetation and fauna – is highly disrupted, fragmented, simplified and species-impoverished in the majority of cities and towns worldwide. Quite often cities and towns arise, even are arising today in many countries, with scant or no consideration given to pre-existing Nature, which makes the starting point for Nature unacceptable. Each city or town, viewed holistically, is at best a very compromised and parasite ecosystem or a miscellaneous loose assemblage of such ecosystems with weak connections between them. There are notable exceptions to this in key areas of some of our cities that have been designed or regenerated with ecological considerations given high priority. However, even the greenest and most environmentally consciously managed cities and towns still do not, in terms of their biotic components and Nature-integrating fabric, fully achieve the status of coherent and complete ecosystems.

PROBLEMS OF A LINEAR METABOLISM

Leaving Nature-based components of our cities to one side, the greatest barrier to regarding cities and towns as 'ecosystems', as some current urban theorists are prone to do, is that most are linear input/output processing systems, referred to in Chapter 1 in the present day's throughput flow of 'take-make-dispose' linear economy, which needs to be changed to a circular flow of 'take-make-reuse-recycle-replenish-reintegrate (back into the natural environment)'.

Products and largely non-renewable energy are taken from 'elsewhere' to go into an urban site (new or existing). The existing model of a town or city as dependent on external and often distant ecosystems is parasitic and damaging, with repercussions at a global scale (e.g. destruction of distant forests cut for wood and associated loss of species and contribution to global climate change). As stated in Chapter 5, natural and semi-natural ecosystems are far more interconnected and interdependent than was once thought, and they exchange components and fluxes of material one with the other. However, these mutualities tend to occur in ways that are not parasitic and which do not harm any of the participating ecosystems.

Manufacturing and consumption occurs during the operational systems of the city or town, where the end products are discarded into the environment once again as if this were an irrelevant 'elsewhere'. The waste gases discharged to the atmosphere cause climate change; the solid waste materials are discarded into the landscape, mostly in landfills from which leachates arise that later pollute our seas and ocean, and polluting liquid wastes are discharged our vital watercourses, water bodies and aquifers. The items discharged and thrown into the oceans include the multitude of plastic products. In many cases (discussed in Chapters 5 and 6), this waste cannot be degraded within even many generational timescales.[10] By contrast, natural and semi-natural ecosystems reprocess and recycle their waste – the outputs from each of their component organisms are used as inputs by one or more of another component. In some of the most efficient systems, like rainforests, hardly any waste is produced at all and no outputs are exported out of the ecosystem.

While existing cities are progressively incorporating higher and higher degrees of recycling initiatives, there is also still huge waste disposal to urban landfill, including the materials in the external and the internal infrastructures from existing cities at the end of their used life. Humanity has come a long way in this regard as a species in recent decades, especially by taxing waste disposal, but there is significantly much more that needs to be done when the urban fabric and artefacts of the human-made world are viewed in their entirety. In summary, the existing contemporary town or city, regarded holistically as a key cumulative human artefact of largely technical infrastructure, does not display many of the normative characteristics of natural or semi-natural ecosystems. There is much to fix, reinvent, redesign and remake in order to make our existing cities, as humanity's largest artefacts, truly ecomimetic.

PROBLEMS WITH SUPPLY OF MATERIALS USED TO MAKE THE TECHNOLOGICAL INFRASTRUCTURE

Are the natural resources we require to create the development of our human-made world likely to be limiting factors – in other words, are we going to run out of them? The concept is often referred to as '**peak supply**' – for example, 'peak oil' and 'peak minerals' (see Glossary). Why is the issue not clear-cut? This is because there are so many assumptions that are made in the analysis of factors left out of the equation altogether.

The first and most obvious constraint is economic cost: because known readily exploitable supplies of oil, for example, are exhausted, countries may be left with relatively low-quality or inaccessible resources for which extraction costs are too high (e.g. deep under the ocean or in the Arctic).

Nevertheless, current literature points out that every time a limiting peak of supply of some mineral or other is predicted, contemporary technology advances so that we either discover a new and previously untapped resource or find a new way to exploit a resource economically when previously doing would have been too expensive. This keeps evolving every time we seem to approach a predicted limit.

However, something more and more observers seem to agree on is that we cannot leave so called 'externalities' out of the equation, most notably the environmental and social harm caused by such exploitation. With the advent of more and more elaborate models inclusive of 'natural capital' and ecosystem service valuations, the net cost of exploiting ever more remote and inaccessible resources is becoming clearer. Even this sort of consideration overlooks the moral obligation to protect other lifeforms with which we share the Planet.

A good example might be the potential exploitation of deep sea thermal vents for their rich associated mineral resources. Initially, this was hoped to be a good environmental option (assuming it could be achieved economically) as there would be nothing living in such harsh conditions. However, this has proven to be a completely inaccurate assumption with the discovery of unique '**extremophile**' (see Glossary) communities of species living very close to the vents at very high temperatures in ecosystems apparently independent of the energy of the sun. These species are not found anywhere else on Earth. So there does not appear to be any exploitation that leaves no damage or environmental and moral scar.

Once we realise that there is a nearly always a trade-off between our material and environmental wealth and health (see Chapter 8), we have choices to make and a balance to strike. It then becomes true that we cannot treat our Planet as an endless supplier of resources for us to make more and more 'stuff' with impunity and with no harm to Nature that we may later regret.

Environmental economists propose a cyclical economy that seeks to decouple humanity's pursuit of increasing societal prosperity from unsustainable environmental impact. This is discussed further in Chapter 8.

PROBLEMS WITH THE FLOW OF MATERIALS IN THE BUILT ENVIRONMENT – DESIGNING TO 'CLOSE THE CYCLE'

For centuries, the ascent of human civilisation has focused on new ways of making things. We have made new compounds previously unknown (or extremely rare) in Nature, in many cases to aid in this manufacture. We have assembled these compounds together into components and artefacts with virtually no consideration of the impact on Earth of the extraction and manufacture processes, and more importantly, no consideration of what happens to the artefacts at the end of their used and useful life. We need to reiterate what was first mentioned in Chapter 1: that the present day's throughput flow of 'take-make-dispose' linear economy needs to be changed to a circular flow of 'take-make-reuse-recycle-replenish-reintegrate (back into the natural environment)'.

Nature does things differently from human society. When Nature makes new things from raw natural ingredients, these are just **assembled materials** – minerals and organic compounds brought together in a temporary combined state before their ultimate decomposition and reabsorption into the environment at the end of their operational life cycles. In some cases, these life cycles may operate on the timescale of millennia (e.g. the weathering of ancient termite mounds), but ultimately, there is a gradual returning of the components to the environment.

Designing to '**close the cycle**' in our ongoing use of materials and artefacts that comprise our human-made world and built environment is an approach that aims to eliminate the very concept of waste. This approach intends to emulate Nature by considering an end fate for all artefacts with an environmentally benign reuse value.

The 'useful life' of a material refers to the entire life cycle of that material, from the point of extraction of materials in their raw state to their making into an artefact to their eventual reconstitution or return back into the natural environment, whether for decomposition or replenishment. The cycle from raw material back to a new raw material for another process is referred to as the material 'life cycle', and the design process that aims to replicate this process is often referred to as 'source-to-source' or 'cradle-to-cradle' (as opposed to the current norm of 'cradle to grave').

A comprehensive Ecological Design process requires a comprehension of the 'ecology' of each and every material and component used. While what happens to the designed system after it has been constructed or manufactured lies outside the control of the designer, it is crucial for the ecological designer to analyse and anticipate all environmental impacts on the biosphere at every life cycle stage and at all geographical scales of reference. Only by doing this can the ecological designer make properly informed decisions on the selection of the materials and processes to be used.

It is also important for the ecological designer to consider the timeframes applicable to each of the impacts and processes carefully. What typically happens today is that future impacts are discounted, as if humanity were trusting in some future technological fix far in the future. The precautionary principle is rarely applied.

The forecasting of possible consequences in these life cycles is, of course, a complex undertaking and very difficult to achieve in a way that is fully inclusive of all environmental, social and economic impacts. But sustainability itself is a concept rooted in care for future generations of humans and other lifeforms, hence forecasting is essential.

As things stand today, most of the existing artefacts, products and materials that we make are not easily reusable or recyclable. Most existing human-made artefacts are inorganic, requiring relatively long timescales of decomposition. In recent years, there has been a surge in scientific interest in estimating (and in some cases testing) how long it takes for different artefacts and their components to biodegrade naturally. What is clear is that the rate of degradation depends very much on the environment in which the artefact ends its useful life. Wet and/or oxygen-rich environments will, of course, degrade artefacts much more quickly than dry and/or anoxic ones (such as landfills).

Naturally occurring organic materials and some human-made organics (like Nylon) and inorganic compounds (like tin) can biodegrade to their component elements or simple compounds thereof within a human lifetime. An aluminium can might take 100 years. By contrast, some compounds that we make, including most plastics, which are long manufactured chains of the same groups of molecules repeated (polymers), often with additional stabilisers, will take much longer timescales than the lifespans of organisms and biological processes present on Earth today to break down into their original organic constituents.

What can happen in exposure to sunlight, heat or strong physical abrasion is that plastics can break up into smaller fragments, called 'microplastics'. Microplastics are dangerous because they are smaller pieces, but they are no more easily compostable than the original artefact. Our seas and oceans are now heavily polluted with these so-called microplastics derived, for example, from larger plastic products breaking up and fibres from synthetic clothing and abrasives used in cleaning products being discharged to the environment in greywater (see Chapter 6).

Microplastics with dimensions of around 10 millionths of a metre (10 microns) are a particular environmental concern. These have been shown to be ingested by marine zooplankton and cause blockages in their alimentary tracts, loss of fat reserves and heightened immune responses – all of which threaten global marine ecosystems by attacking the base of the marine food web. Additionally, microplastics accumulate up the food chain into the seafood that winds up on our plates. Plastic fibres have been found to contaminate 84% of drinking water samples worldwide, thus becoming a public health concern. It is choking both our waterways and our bloodstreams!

Closing the loop of products and chemical components in what we make is not just about recycling, but quite fundamentally about **substitution** – finding environmentally benign alternatives to the potentially environmentally harmful ingredients we currently may be using. This is difficult for many reasons, as the substitute needs to function at least as well as what it is replacing, and that means the available options have many similar characteristics and potential for environmental harm to the ingredients for which substitutes may be sought.

Addressing comprehensively the cyclic ecological logistics of materials in the human-made world is an ambitious and massive proposition that will not be achievable overnight. Yet if humanity does not persevere in this, our Planet will become increasingly littered with used materials and products accumulating in its terrestrial, aquatic and aerial environments. We will then continue to perpetuate the concept 'waste', and suffer its harmful effects, rather than take the crucial lessons from Nature and cease providing ourselves with huge and often intractable problems to solve.

AVOIDING PIECEMEAL AND INCREMENTAL APPROACHES

Part of designing to close the loop involves designing truly holistically and avoiding incremental approaches. Chapter 4, in discussion of the need for an infrastructural approach, explained why the redesigning and remaking of the built environment must avoid incremental approaches, producing components that appear to deliver ecologically sustainable outcomes, but in fact do not because they depend on other systems that are themselves not ecologically designed. To undertake design in a way that both incorporates proper analysis of component life cycles and ensures holistic integration of all relevant elements is a huge logistical and analytical challenge. This, then, is another crucial challenge for Ecological Design.

HARNESSING THE FOURTH INDUSTRIAL REVOLUTION FOR ENVIRONMENTAL SUSTAINABILITY

There have been several peaks of activity in the development of new technological approaches by humankind. These are often referred to as 'industrial revolutions', the first three beginning in the 18th century, when significant proportions of the population of Europe and North America moved from rural lifestyles to urban ones dominated by the steam-driven iron and textile industries. The Second Industrial Revolution, just before the First World War, saw the emergence of mass production driven by electrical power, and development of the first telecommunications technologies. The Third (or 'Digital') Revolution, starting around the 1980s, saw the main change from analogue to digital technologies and the true emergence of information and computer technologies served by computers and the internet.

The progression of technology and engineering in this way has led to ever greater and more complex environmental side effects. The Third Industrial Revolution in particular has been associated with high levels of damage to the environment from the extraction, manufacturing and end-use disposal of ever more sophisticated electronic components relying on suites of notably toxic metals such as lead and '**hexavalent chromium**' (see Glossary).

Our engineering infrastructures that are generally and currently environmentally harmful need to be rethought, redesigned, reinvented and remade so that they seamlessly biointegrate and connect with Nature to restore rather than harm it.

However, part of the solution, ironically, seems to lie with new technology that integrates with Nature to provide higher processing capacity sufficient to start fully understanding systems as complex as the atmosphere, for example. In 2015, Klaus Schwab of the World Economic Forum introduced the concept of the **Fourth Industrial Revolution** as one of huge advances in the way technology can be integrated with Nature and natural systems and human society, including human beings themselves. Key components of this approach (see Glossary for a short definition of each) include applications of: **nanotechnology, biotechnology, quantum computing, artificial intelligence**, the **Internet of Things, 3D printing** and **autonomous vehicles.**

The applications and potential of this for our redesigned world in terms of long-term sustainability are enormous. We may be able to far more reliably predict and avoid environmental harm, increase our ability to address challenges of infrastructure far more efficiently, avoiding problems before they arise and generally hugely improve our efficiency in use of resources.

When we talk about integration of infrastructures (see Chapter 5), the Fourth Industrial Revolution will be key to the seamless biointegration of the biotic and abiotic to create benign integrated novel urban ecosystems that are restorative of the biosphere and the operation of which can be optimised. To see whether or not we are succeeding in this regard, we must use all of these new technologies to enable a global scale of environmental surveillance; and we are now indeed starting to do just that.

REAL-TIME GLOBAL ENVIRONMENTAL SURVEILLANCE

A crucial aspect of Ecological Design is the need to globally monitor, measure, assess, evaluate, comprehend, forecast and then mitigate the environmental effects of our actions. We need good monitoring data that is real-time and continuous. This needs to be applied to designed systems at each step of their materials' logistics life-cycles and then summate their effects. The effort has to be dynamic and constantly adaptable, not static.

For the first time, with advances in integrative and information technologies, we have the potential opportunity of moving towards such real-time sophisticated monitoring of global 'spheres' and integrating the information from sensors into data that enables us to respond adaptively and help maintain the Earth within its environmental limits for life support.

REMAKING AND REINVENTING THE TECHNOLOGICAL INFRASTRUCTURE

In summary, this chapter has addressed human society's technological infrastructure. In order that this can be benignly reintegrated with the other infrastructures, being the Nature-based, the hydrological and the socio-economic-political infrastructures.

It is clear that there is much work to be done by humanity with the technological infrastructure, as stated in Chapter 1. For a resilient, durable and sustainable future for human society, we need to repurpose, reinvent, redesign, remake and recover our human-made world so that our built environment and all the artefacts and structures are seamlessly biointegrated with Nature and function synergistically with it. We are far from achieving this state, but this is another great challenge of Ecological Design, and humanity must take concerted action to remedy the current state.

NOTES

1 Jones, C. G., et al. 'Organisms as Ecosystem Engineers.' *Oikos*, vol. 69, no. 3, 1994, pp. 373–386. JSTOR: www.jstor.org/stable/3545850.

2 Berner, R. A. *The Phanerozoic Carbon Cycle: CO_2 and O_2*. Oxford University Press, 2004.

3 Environmental Literacy Council. 'Biogeochemical Cycles.' 2018, https://enviroliteracy.org/air-climate-weather/biogeochemical-cycles/.

4 Cain, M. L., et al. *Ecology*. Sinauer Associates, Inc. Publishers, 2014 (Ch. 22).

5 Collias, N. E., and Collias, E. C. *Nest Building and Bird Behavior*, vol. 857. Princeton University Press, 2016.

6 Ford, T. E., and Naiman, R. J. 'Alteration of Carbon Cycling by Beaver: Methane Evasion Rates from Boreal Forest Streams and Rivers.' *Canadian Journal of Zoology*, vol. 66, no. 2, 1988, pp. 529–533. doi:10.1139/z88-076.

7 Funch, R. R. 'Termite Mounds as Dominant Land Forms in Semiarid Northeastern Brazil.' *Journal of Arid Environments*, vol. 122, 2015, pp. 27–29. doi:10.1016/j.jaridenv.2015.05.010.

8 Sorokin, Y. I. *Coral Reef Ecology*, vol. 102. Springer, 1995.

9 Chappell, J. 'Coral Morphology, Diversity and Reef Growth.' *Nature*, vol. 286, no. 5770, 1980, pp. 249–252. doi:10.1038/286249a0.

10 Zalasiewicz, J., et al. 'The Geological Cycle of Plastics and Their Use as a Stratigraphic Indicator of the Anthropocene.' *Anthropocene*, vol. 13, 2016, pp. 4–17. doi:10.1016/j.ancene.2016.01.002.

FURTHER READING

Braungart, M. & McDonough, W. 2009. *Cradle to Cradle. Remaking the Way We Make Things.* Vintage, New York.

Power, A. 2000. *Cities for a Small Country: The Future of Cities.* Faber & Faber, London.

Rogers, R. 1997. *Cities for a Small Planet.* Faber & Faber, London.

TUCK

8

Anthropocentric infrastructure

HUMAN SOCIETY AND SOCIAL-ECONOMIC-POLITICAL SYSTEMS AS INFRASTRUCTURE

A contention in Ecological Design is that human society and its societal systems must themselves be regarded as an infrastructure that needs to be similarly biointegrated with other infrastructures to make our quatrobrid constructed ecosystem. It is clear that in most approaches to Ecological Design, inadequate attention is given to societal aspects and their crucial influence on ecological outcomes. Human ideology determines the societal systems that humanity creates for itself, including social-economic-political-institutional-educational systems. These societal systems determine the physical system of the built environment and its multitude of artefacts.

In addressing the current issues of environmental harm and striving to work within environmental limits while meeting the needs of the global human population, we need to also address the 'virtual systems' that run and emanate from human society. Accordingly, as set out in Chapter 4, the fourth key element of Ecological Design is consideration of the 'human factor'. This takes the form of our social, economic, political, religious, institutional and legal systems, together with the actions and interactions of all of humankind – the mutualities of interest and trust that are the strength of human communities and social connections. We need to point out that this factor is often missing or omitted in the sustainable design approach of many designers.

The 20th-century French philosopher Felix Guattari saw that human socio-economic-political systems and processes have their own form of 'virtual ecology'. He went further and pointed out that the *ideas and belief-sets* held by people have a 'life' and hence an 'ecology' of their own, often with emergent properties that are very hard to predict even with extremely detailed knowledge of society and politics. He further spoke of 'three ecologies': the ecology of Nature, the ecology of human interactions and the ecology of ideas.

This combination of humanities systems, organisations, rules, ideologies and ideas can be thought of as a part-physical, part-virtual infrastructure that is referred to here as '**anthropocentric infrastructure**' (see Glossary), as it is completely related to the existence and interests of humankind.

How we treat Nature and how we use technology are completely determined by the emergent properties of the complex interactions within this infrastructure. Many existing approaches to green design poorly consider this infrastructure, especially in the creation and operation of the built environment. Examples of this are inqdeauqtely considered in many international accreditation systems (such as LEED or BREEAM: see the discussion of these in Chapter 2).

Natural phenomena in the form of volcanic eruptions or axial wobbles of the Earth, can, of course, cause major environmental changes and extinction of species. In the present Anthropocene era, however, the most real and present danger to human society and life on Earth is anthropogenic, and relates to the progressive degradation of the global environment.

The physical and constructed systems on the Planet are created by humans and are directed by their social-economic-political-institutional systems. Most of these systems have been created and built without due consideration of their negative impacts on Nature, and have not been driven and guided by the science of ecology. This is the main cause of the current state of the Planet's environmental disrepair.

Any effective action to mitigate and eliminate environmental degradation, then, starts with humanity: so despite its arcane complexity, anthropocentric infrastructure must be addressed head-on in Ecological Design. In other words, Ecological Design needs to include the endeavour of addressing the reexamination of human ideologies and the transformation of human societal systems to focus on Nature, on the biosphere and on the Planet's biogeochemical cycles.

ANTHROPOCENTRIC INFRASTRUCTURE'S CONTROL OF OTHER INFRASTRUCTURES

Human services and the work of human society are necessary, of course, for the coordinated management, efficient functioning, regular maintenance and constant monitoring of all technological infrastructure. While our technological infrastructure can operate in part by intelligent ('smart', automated or artificial intelligence systems, these systems (for now, at least!) require human direction, programming, overseeing, supporting servicing and monitoring. It is the human infrastructure that makes the technological and engineering systems work, no matter how automated. There are, of course, science fiction scenarios in which artificial intelligence (AI) becomes effectively self-aware. Indeed, there are those who would argue that only by the full hybridisation of humans with technology will we ensure the ongoing pre-eminence of humankind.

Furthermore, given the power of human intervention and the effect of human predominance on Earth today, humanity now also plays a major role in the regulation of the hydrologic cycle (beyond urban water reticulation and supply systems). Similarly, for good or ill, humankind now has significant dominion over Nature-based infrastructure. Hence the obligation, both practical and moral, falls upon Earth's anthropocentric infrastructure to ensure the healthy state of the Planet's ecosystems and biogeochemical cycles: in other words, making sure that Nature's life support systems are functioning effectively for all of global biodiversity.

These sorts of considerations raise profound ideological and ethical issues. However, before we consider these issues and the challenge of changing hearts and minds towards an ecocentric approach, let us first examine what is wrong with much of modern-day anthropological infrastructure by considering our currently prevailing economic systems.

TRADITIONAL ECONOMIC APPROACH WITH ENVIRONMENTAL ISSUES EXTERNALISED

Much of the world still operates with virtually zero integration of environmental impacts or other sustainability criteria within economic models. Citizens are seen as providing labour to corporations and governments which in turn provide goods and services to citizens in a never-ending economic loop. Clearly, the problem with this model is that it ignores the dependence of the whole system on the environment and its resource limitations. It also overlooks issues such as social equity and fairness across the Planet.

LINEAR VERSUS CIRCULAR ECONOMIES

Due to lack of public consideration for the environment, the world still operates on the basis of what is often referred to as a '**linear economy**' (see Glossary). This is an economic model based on the 'extract, make, use, discard' approach to the operation of our human-made world.

An alternative concept that has developed in recent decades is that of a '**circular economy**' (see Glossary). In such an economy, inputs and outputs are cyclical: one output (previously considered waste) becomes the input for another sustainable process. In this way, the life-cycles of materials resemble loops rather than lines. The process is never fully complete. There are always leakages in any real system, but the aim is to slow, narrow and eventually close as many of them as possible.

Practices associated with the circular economic approach include recycling (including refurbishing and upcycling) – an approach that is becoming more and more commonly adopted worldwide. Clearly the mindset change required to convert human society from operating linear economies to embracing circular ones needs to be quite profound. There needs to be a new focus on 'systems thinking'.

Part of the circular economic approach also entails making products last longer and work more efficiently, rather than constantly producing items designed to be used for a short period of time. We need to turn our creative energies and mindsets to challenges of product life extension and efficiency too. This exists in sharp contrast to the insanity of the linear economy, in which redundancy is built in to many consumables to maintain ongoing sales.

Crucial to successful implementation of circular economic approaches is to do so while maintaining revenue generation. By emphasising efficiency (and eventually, as the approaches become standardised, economies of scale) and focusing on quality rather than quantity, circular economic approaches can be profitably viable. However, to secure this outcome, support from political and financial institutions is necessary in the form of subsidies (that destructive products, such as fossil fuels and industrial agriculture currently benefit from) and investment.

Another fundamental concept of the circular economy, is that it can lead to a higher, not lower, quality of life for the consumer. It is also important to note that the circular economy is a concept that applies regardless of the socio-economic status of the place or culture adopting the approach. Of course, there is a strong natural analogy to ecosystems that work on cyclical metabolic processes. This is a core element of ecomimicry.

DECOUPLING ECONOMIC GROWTH FROM EVER-INCREASING RESOURCE USE

Another key benefit of the circular economy approach that deserves its own sub-heading here is how we address our ever-increasing use of global resources, both renewable and non-renewable. We should, and in some cases now do, legislate for or incentivise the production of more efficient artefacts that use less inputs, especially energy (a good example being the efficiency labelling of electrical devices in the home, so-called 'white goods').

Another method, is to directly encourage society to voluntarily adopt a more frugal approach to resource use through convincing moral argument (see later in this chapter).

'Doing more with less', sometimes referred to as 'sustainable retreat', must be portrayed by leaders and governments as a step forward and a progressive approach. To convince significant proportions of society to convert to this sort of thinking requires efforts to combat the tendency for such approaches to be considered somehow retrograde and regressive.

DECOUPLING ECONOMIC GROWTH FROM THE USE OF HARMFUL SUBSTANCES AND MATERIALS

To take an ecologically sound approach to economics, another key required approach is to move ever closer towards a decoupling of continued economic growth from the continued and often expanding use of extracted or manufactured substances and materials that pose a significant risk to the environment.

As discussed in previous chapters, a compound that is fully safe when incorporated into an operational artefact may become extremely hazardous when released accidentally (or carelessly discarded as waste). One approach is to reduce the increased use by ensuring any such materials are benignly recycled for reuse (see 'Linear versus circular economies' later in this chapter). But it is even better if we encourage the discovery and use of environmentally benign substitutes that pose no environmental hazard. This can be done by, for example, government directives sponsorship of research, by incentives and taxes on industry, and by changing requirements for labelling and advertising that may help change consumer purchasing habits.

HUMANITY TO ADOPT A SUSTAINABLE DIET

A key element of anthropologic infrastructure is the highly complex and culturally diverse pattern of food consumption in human society. After reaching a point of survival, the politics and framework for food production were driven largely by desire to obtain new and interesting tastes and to eat more and more of the foods that in the earliest stages of our evolution we would have enjoyed only rarely (such as red meat). The human infrastructure of food provision includes some of the world's strongest socio-political lobbies, such as commercial fisheries and commercial farming industries.

Traditionally, communities have managed to consume resources well within their natural capacity to be replenished, and some communities still maintain this low-impact lifestyle. However, modern systems of food production, such as industrial animal husbandry, reach far beyond natural supply and exploit ecosystems well past recovery.

The absurdity of the modern food system is that it promotes such large-scale production of food that is unhealthy and causes diet-related disease, such as obesity, heart disease and diabetes, which follow closely behind 'development' in every country! Countries that until recently never had some of these ailments thanks to a traditionally largely vegetarian diet are now succumbing to the same suite of ailments. Moreover, production of luxury items in the developing world for consumption in the developed world actually robs the developing world of desperately needed land for sustainable growth of staple crops.

The ecological cost, and hence economic impact, through loss of ecosystem services is enormous. Examples already mentioned have included:

- Loss of natural habitats to new crop or forestry production areas, huge carbon emissions from tillage, food transport and livestock.
- Pollution on a global scale from agricultural fertilisers, herbicides and pesticides.
- Profligate consumption of freshwater resources.
- The contingent loss, in all cases, of biodiversity.

At the same time, we need to be aware that the Planet is becoming too small for human communities where global population is increasing at an alarming rate and humanity is in danger of self-destructing. Global food security is an issue in urgent need of solutions. Humanity and its built environment need to actively pursue alternative ways of living, diets and functioning if humanity is to have a sustainable future to survive indefinitely. The issue of food production is a good example of how the privileged developed world has used and abused natural systems and resources to the disadvantage of the underprivileged and developing segments of human society, exacerbating the present environmental crisis. This issue is destabilising in a host of ways that threaten the very fabric of anthropological infrastructure, for example through mass emigration from developing to developed countries.

DESIGNING FOR HUMANITY'S HEALTH, WELL-BEING AND HAPPINESS

Using GDP as the measure of economic growth is fundamentally problematic as it self-perpetuates the myth that actual net income is the only viable societal goal.

Human physical and psychological health are generally externalised from the models used, along even with more abstract concepts, such as human happiness. We therefore arrive at the bizarre situation so much discussed by philosophers and pundits of 'losing as we gain'. In other words, we are living in an ever more frenzied world of capitalist acquisition of material wealth driven by mega-scale production and highly sophisticated marketing efforts that addicts humanity to want 'more and more' goods that, following an initial increase in human satisfaction, is characterised by minimal increases in human contentment with further acquisitions, and eventually actual net declines. The result of this crazed drive is generally to ignore the harm it does, not only to the environment, but to ourselves.

If we are to avoid the illogicality and addiction of the so called **acquisitive culture**, the norms and values of most of human society that link net financial profit to human progress, and which form the backbone of the anthropocentric infrastructure through much of the world, need to be changed. There needs to be a mental decoupling between our measurement of economic growth and our concept of human holistic well-being. Put another way, human society must change its perception of what it needs to enjoy a high **quality of life** and focus on that, rather than solely on **material standard of living**. This is the only way we will stem our ever-increasing and unsustainable pattern of use in the consumption of materials and natural resources.

In effect, we need to evolve ecology-based social-economic-political systems to encourage humanity to voluntarily select a way of life that adequately meets the needs of civilisation, and does not render us hopeless addicts to meaningless, profligate, wasteful and environmentally harmful continuous acquisition.

It is worth noting that the harm done engages in a positive feedback loop: the system fuels demand, then demand fuels the system to provide ever more artefacts, consume ever more energy and discharge ever more waste and environmental pollution. Combined with human population growth, these magnifying effects pose a real hazard to the biosphere. In economic terms, we need sustainable 'demand management', where society's sustainable procurement is matched by sensible and frugal use of natural capital by a happier humanity and a precautionary approach to its sustainable preservation.

DESIGNING FOR SOCIAL EQUITY

It is obviously true that development and technology have enhanced the standard of living and well-being of many people in the world. However, this has been achieved at a great and unsustainable environmental cost. There is now a pressing need for transformation of human culture in both the developed and developing world to one that is far less materially acquisitive, so that available resources can be more fairly distributed, thereby avoiding globally destabilising consequences.

ADDRESSING HUMANITARIAN ISSUES

Of course, humanitarian concerns extend far beyond food. The world is still greatly afflicted by humanitarian issues such as poverty, disease, unemployment, human trafficking, society-destroying drug addiction, organised crime, civil conflict and planned control of human population growth. The purview of the designer in reimagining and remaking the human-made world so that it becomes ecologically effective and ecologically benign in its manufacture, operation and disposal processes has a potentially huge effect on all of these humanitarian concerns.

Of course, it is for administrations and governments to set policy that directly addresses these issues. But so many of these ills stem from inadequate access to or conflict over essential resources. There are already wars over water. There are conflicts over exported pollution and waste. The stress of the threat of international conflagration helps drive a search for psychological numbing that is readily provided by the use of drugs.

In particular, with the challenges of each generation come the insecurities and concerns of the young. Those who grew up in the Cold War will remember the occasional thought of 'What's the point?' as humanity was about to send itself into oblivion at any moment. That particular fear is perhaps increasing again today.

However, as stated previously, the most real and present (urgent) danger is the harm we are doing to our own life support system, progressively and inexorably, by failure to adopt an ecocentric approach to life. The lyrics of young poets and songwriters today are charged with laments for the depredations we are wreaking on the environment. There is so much news about the Planet, and none of it seems good. This is highly disempowering and helps to fuel socio-pathologies in our youth across the world.

By changing to an ecocentric approach to design our human-made world, we simultaneously reduce the likelihood of so many other social ills and challenges from gaining a further foothold and flourishing at the expense of society, our human civilisations and the Planet.

VARIATIONS OF 'SUSTAINABLE ECONOMICS'

Efforts have been made in recent decades to develop new economic approaches that try to address the sustainability equation. In industry, commerce and sometimes in politics too, monikers are often used to summarise the approach succinctly. One example of this is a reference frequently made to 'sustainable' business respecting the '**triple bottom line**' –the set of objectives that consists of 'social equity, economics and environment'. Another example is the **three pillars of 'People, Profit, Planet'** referred to in the international conferences of the United Nations.

Another conceptual economic model for sustainability is that of so-called '**Doughnut Economics**'. This model is actually quite a useful one conceptually in that it demonstrates the idea of limits visually quite well. Sustainable economic activity is seen as taking place within the main body of a two-dimensional doughnut-shaped economy. The outside face of the doughnut represents the natural resources of the Earth and the limits these impose on traditional economic growth. The inner ring represents the minimum degree of development required to eliminate poverty and ensure that the global human population has a reasonable standard of living, free of the worst social pathologies that are increasing in many areas today.

All of these mantras are useful because they bring more of the overall sustainability equation to the forefront of economic thinking. However, none is comprehensive. All omit explicit reference to other factors such as political and cultural aspects. Technology and its consequences for good and ill are not fully considered. The objectives set within these sustainability systems are often subject to multiple interpretations, and it falls to the party using them to decide where the emphasis should lie between the factors actually being considered.

The ecological imperative and Nature-centric approach asserts that the natural environment must *come first* because endeavours to address society's socio-economic-political-physical systems become easier if we have a clean environment – clean land, clean air, clean water, clean soil, clean ecosystems – within which society itself has a clean bill of health, but this priority may not be the case with other proponents. It also asserts that *all* elements of all infrastructures (Nature-based, hydrological, technological and anthropocentric) must be considered together, with nothing left out and all fully integrated one with the other, if we are going to ensure the resulting systems endure and thrive.

PROBLEMS WITH MONETISING NATURAL CAPITAL AND ECOSYSTEM SERVICES

Putting some economic value on the environment is certainly a better approach than not caring about the environment at all. In formal economic language, this is about internalising into economic and financial models the 'externalities' of the value of, and impacts on, the natural environment. However, it is also problematic when taken too far. Indeed, many analysts, economists and even some ecologists today seek to justify ecological and sustainable approaches purely on the basis of macroeconomic arguments. The suggestion is that, once we have monetised natural capital and ecosystem services correctly ('got the right price for the environment'), the market will always come up with the most efficient and fair apportioning of the responsibility for shouldering, avoiding or addressing the environmental harm caused by human activity, and to 'keep the Planet healthy'.

In specific cases, such approaches may be partly justified if, within a given socio-political regime, they do result in stemming of the tide of environmental destruction or indeed aiding environmental restoration. It could be argued that to some extent, the global impact of greenhouse gas emissions between countries creates an equitable share of the responsibility for stemming global climate change between more polluting and less polluting nations.

There are, however, several key problems with such approaches.

The first is that no inclusive economic model can ever be *completely* inclusive of all the effects on the natural environment, given the huge complexity and interrelationship of ecosystems and natural global cycles.

Secondly, there are many ways to assign economic value, which are all flawed by being rooted in human ideology and societal systems (such as 'willingness to pay'), and these methods can produce very different results and conclusions.

Thirdly, markets are volatile. Trying to put a value on some aspect of the environment as it is relevant to society today could quickly alter in the future when the relative value has dramatically changed.

Fourthly, most models depend on random, rational and fair competition between small economic entities that keep their word. This does not allow room for human nature. It does not, for example, allow for the formation of monopolies or cartels and the development of price fixing, for bribery and corruption that can lead to false verification and auditing of environmental markets.

Fifthly, current macroeconomic models relating to environmental goods, services and harm treat these tradable items as they would any other commodity, with no further investment. But there is something special about the global environment: we only have one, we rely on it, and we cannot afford to damage it! We cannot afford to get things wrong in the trial-and-error system of the market as we might if we were selling pencils or ball bearings.

Finally, and most importantly, the ecocentric ideology for humanity is ultimately a matter of **ethics** – that is to say, dealing with **human ideologies** so that we 'do the right thing'. This is perhaps a fundamental case of the first problem listed, that we cannot price everything because some things are inherently priceless.

Holistic Ecological Design does not merely aim at ensuring humanity's long-term survival and the flourishing of human civilisation, but has the parallel goal of sustaining all lifeforms and habitats on the Planet. The weight of this *moral* obligation towards life on Earth stems not only from humanity's special ability to wreak environmental harm, but also our unique ability to analyse, predict and take action to restore the global environment.

Besides the infrastructures above, Ecological Design needs to consider humanity's ideologies and how we might change them, and to consider human society's social, cultural, economic, political, institutional and educational systems as infrastructure. This is an aspect of Ecological Design that is commonly neglected or omitted by designers.

HUMAN IDEOLOGIES AND ECOCENTRICITY

The term 'ideology' was conceived as the 'science of ideas'. Humanity's ideologies are human society's mindsets, defined here as collections of normative beliefs and values that an individual or group holds for reasons other than those based in simple matters of fact (so-called epistemic reasons). Ideologies have been defined as *'the imagined existence (or idea) of things as it relates to the real conditions of existence'*. They arise from how we philosophically and emotionally, as well as rationally, develop concepts, and these in turn determine our cultures and ways of life. There are a vast array of human ideologies across the Planet, which until the arrival of the global internet emerged quite often in relative isolation one from the other.

As stated earlier, our great array of societal socio-economic-political-institutional systems determines the physical systems and artefacts that are made and how they are constructed, used and operated. It may seem obvious to state, but it is worth emphasising, that as a species we have unusually wide choices in how we operate our lives, especially those of us living in the developed world.

There are choices to be made about the energy systems we use to power human society and build structures and infrastructures, urban areas and whole cities. We can choose how people and goods are moved and transported. We can choose what we should eat and how we should produce the food. We can choose the means by which the food gets to our table, including the distance the food has to travel. We can choose between different types of semi-natural production systems such as agriculture, aquaculture and forestry. If we do not make these choices with due deference to Nature and the science of ecology, with the power of increasing population on Earth and rapidly developing technology, we will have the capacity for enormous negative environmental impacts at the Planetary scale.

The premise of this chapter is that our varying ideologies need to be realigned regardless of social and cultural differences, or the different echelons of society, so as to be ecocentric rather than technocentric or anthropocentric.

The realignment may involve a certain amount of cultural homogenisation, which is often resisted in the interests of diversity and creativity. However, not all diversity is good if it includes approaches that destroy the biosphere and disrupt whole Planetary cycles. Alignment on ecocentricity is therefore essential if we are to address global environment impairment and threats effectively.

THE IDEOLOGY OF ECOCENTRICITY

Ecocentric or ecology-centric ideology directs us to not be exploitative of Nature, and not to negate or ignore the intricacy and interdependence of Nature's systems and all of its lifeforms. Humanity must develop an ideology that not only recognises the right of other species to share the Earth with us, but also fully comprehends and protects the ability of Nature to deliver ecosystem services. Humanity must adopt an ideology that directs us to avoid the continued callous clearance of the Planet's natural and semi-natural habitats without consideration of all local and global ecological consequences. Humanity must integrate into this a clear understanding that we can no longer treat Nature's terrestrial, aquatic and aerial environments as sinks for humanity's wastes and emissions.

We need to put in place an enduring integrated, symbiotic stewardship and partnership with Nature, working within its thresholds of resilience (see Chapter 5) and acting with precaution and without profligacy.

In practical terms, adopting an ecocentric ideology means that with each development or restoration project, we move closer to being true stewards of the Planet's biodiversity, systems and natural resources, and away from a condition in which we become presumptuous and voracious exploiters and despoilers of the web of life of which we form but part.

The move to true ecocentric sustainability is a challenge from the political to the economic, to the commercial corporate, to the industrial sectors, to international and national, legal and legislative systems, to human communities and to the individual. Changing society's social-economic-political-institutional systems has to be effected both by local political communities and by global institutions.

All of humanity must collectively take a share in what is a global responsibility to end Earth's current environmental predicament, part of which is inherited as a result of actions of our forebears and part of which relates to harm we are bequeathing to future generations of all species. The transformation of humanity's ideologies and its behaviour to be ecocentric and ecology-driven needs to be achieved through public persuasion and education, the endeavours of behaviour change-makers, determined leadership and political will.

Although we are in one sense just a species in Nature, it should be clear by now that we have an unprecedented ability to transform the global environment and decide whether other species thrive or perish. Different ideologies around the world vary in their approaches to the moral stewardship of Nature. Unfortunately, too often the stewardship in itself is anthropocentric – aimed at managing Nature more or less solely for the benefit of human beings and human society and civilisations. Such an anthropocentric approach, however, will not achieve sustainability: those societies and civilisations place too little weight and consideration on the rest of the biosphere and Nature's life support systems that depend on global biodiversity (see Chapter 5).

All of the many and varied insults to Nature, which have already been described in this book and which we bring about as a byproduct of our ideological anthropocentricity, have occurred and are occurring because of the incomplete and inadequate nature of our prevailing ideologies, to the point where we ourselves and our societal systems and civilisations are seriously threatened.

Ideological realignment towards ecocentricity is, then, the crucial imperative and challenge we must meet if we are to redesign and remake the human-made world.

TRANSFORMING HUMAN IDEOLOGY AND ETHICS

What power do Ecological Design and its practitioners have to effect the changes in humanity's existing ideology and ethical frameworks? It would seem logical to state that such change can only be effected by the leaders (political, religious and cultural) in human society around the world. But this is to ignore the ecology of ideas. Every incremental step that is made to explain the environmental dilemma, how a design or human intervention has been contrived to avoid net environmental harm and, where possible, environmentally repair and restore can plant an ideological seed in the mind of anyone who comes into contact with it physically or virtually. In effect, the success of Ecological Design can be achieved by thought leadership and by designing innovative prototypes.

Informed Ecological Designers can in this way lead by example, and the cumulative effect of the interventions and explanations in increasing the ecological literacy of society (see later in this chapter) may have unexpected consequences. It can lead, for example, to the unexpected appearance of environmental pressure groups and movements, sometimes given power overnight by the internet and social media, that can then place enormous pressure on industry and governments alike to change their approach.

In an extreme case, it can lead to changes in administration and governments in favour of leaders who are prepared to apply a truly ecocentric approach in the making of law and policy.

ECOLOGICAL LITERACY

When human society has no sense of the complexity, interrelation or responsibility for ecological systems, it cannot be expected to act in defence of Nature. It is also crucial for all involved in the design process (particularly those whose work directly and immediately impinges on the natural environment, including designers, architects, planners, engineers, builders, environmental policy makers and others in the commercial and real estate business) to possess ecological literacy in order to be able to effect environmentally sustainable Ecological Design.

Much is said in the modern environmental literature about the process of alienation that has occurred between human society and Nature. Some would argue that the blame for this falls largely on modernist design ideology and how it places a transcendent importance on humankind and our right of dominion over Nature. We have in recent years become a predominantly urban species which neither understands Nature nor comprehends the degree of our dependence upon it. In many of our urban areas around the world, what is left of Nature is highly contrived or downgraded, and there is often a brutal divide between the urban and rural realms. The urban citizen has no daily experience of complex, natural processes and phenomena to maintain an awareness and foster understanding of them. With some notable exceptions, when efforts are made to counteract these overwhelming trends by environmental educators and concerned citizens, the approach taken is to remove the student from the urban realm to experience Nature in the least human-modified area of countryside that can be reached economically from the student's place of residence. Acquiring ecological literacy involves engaging in 'unlearning' our current approach to Nature and relearning all over again a brand new ecocentric one. It calls for a totally fresh mindset, and indeed, seeing the global environment and human society through new eyes, 'ecocentrically'. More importantly, it makes evident the need to revise humanity's existing social-economic-political systems that significantly control the fate of the global environment.

Acquiring ecological literacy should further make evident to human society the inadequacy of current approaches to designing and making – where so often ecological considerations are absent, ignored or even denied – and the need to devise and adopt new design norms and approaches.

A further usefulness in humanity acquiring a literacy in ecology is the engagement of human society with the concept and noble endeavour of restoration ecology, and how hard it is to restore natural and semi-natural ecosystems after they have become significantly damaged.

Accomplishing this literacy is a daunting task as it covers a very large body of literature that covers ecology, humanities, engineering and technology, hydrology and other related disciplines.

THE ANTHROPOCENTRIC INFRASTRUCTURE

All humans and the multitude of socio-political, institutional, legislative and ideological systems of humankind make up, then, what may be seen as our '**anthropocentric infrastructure**'. Proper consideration of this is rarely addressed in projects purporting to be sustainable. By this omission, designers are missing a key element in the enduring impact and uptake of Ecological Design.

Every effort should be made to help convert humankind to a new understanding and awareness, a new ideology that does not regard Nature as an infinite source of natural resources to be consumed and exploited. It must be cognisant of Nature's ecological systems, the intricate web of life and the ecosystem services it provides for all life without our intervention. The ideology must not regard the Planet's landscapes as land to be callously cleared of habitats, altered and levelled without consideration of the ecological consequences. It must prevent us from treating Nature's terrestrial, aquatic and aerial environments as sinks for humanity's wastes and emissions.

In a rigorous and comprehensive Ecological Design approach, the **anthropological infrastructure** needs to be combined and biointegrated together with the other infrastructures in design – the **Nature-based infrastructure**, the **technological infrastructure** and the h**ydrological infrastructure** – in an integrated matrix of infrastructures which enables proper coordinated functioning of the interactions, synergies, conflicts and emergent properties. The emergent designed system, as a quatrobrid constructed ecosystem, must be virtual as well as physical, with resonance far exceeding a given place or time of intervention.

FURTHER READING

Biodiversity by Design & Yeang, K. 2011. *Designing for Biodiversity Productivity and Profit.* British Council for Offices, London.

CABE Space. 2003. *Does Money Grow on Trees?* CABE, London.

Jackson, T. 2011. *Prosperity without Growth: Economics for a Finite Planet.* Routledge, London.

Papanek, V. 1995. *The Green Imperative: Ecology and Ethics in Design and Architecture.* Thames & Hudson, London.

Terrapin Bright Green. 2012. *The Economics of Biophilia: Why Designing with Nature in Mind Makes Economic Sense.* Terrapin Bright Green, New York.

Wines, J. 2000. *Green Architecture.* Taschen, Cologne, Germany.

9

Being 'at one' with Nature

OUR EFFECTS ON PLANET EARTH

The end objective of Ecological Design is clear – to achieve a state where human communities and their built artefacts are seamlessly and benignly **at one with Nature.** In reaching this point, an understanding should have been reached that there is an overriding issue for human society to address in terms of the extensive state of devastation of the global environment. Chapter 1 set out a brief precis of this global condition.

All of human society needs to recognise that without concerted, systematic, cooperative and sustained effort by humankind, we are fast approaching the point where we have put ourselves on the ever-growing list of species at threat of extinction. If humans do not become agents of our own extinction, then at the least we can expect a global conflagration on a terrible and massive scale over ever-diminishing resources and the 'services' previously provided by Nature that we are destroying in a wholesale and reckless fashion worldwide.

Some less forgiving sectors of our human society might assert that the demise of our human species would be a good thing. The literature has many contributors who describe *Homo sapiens* as an organism that has gone out of control on Earth and is pushing natural systems beyond their limits of resilience through a combination of systemic weakness, short-term greed and corruption. These are hardly the credentials of a 'wise man' (strictly translating the Latin name for our species, *Homo sapiens*). If we were to render the Planet unsuitable to support us (or most of us), then this might give a new lease of life to many other lifeforms. Perhaps the cephalopods (octopuses, squids and relatives), which independently evolved an astonishing level of intelligence, might one day rise to the level of global ecosystem manipulators, and make a better job of it than we have done!

However, the great belief here is in the value and best qualities of humankind. The human species is, after all, the most extraordinary species known to science. Humans are capable of the sublime and extraordinary as much as the unpleasant and destructive. We need urgently to change the balance, tenet and direction of our relationship with the rest of the Planet.

It would be comforting to hope that mere accumulation of material wealth and good economic status will make us better Planetary stewards – and hence, by focusing merely on raising people out of poverty, we would eventually resolve the Earth's sustainability equation. Of course, economic development has had major benefits for many people on Earth, but at huge cost to the natural environment. Moreover, we seem, in developed and undeveloped countries alike, so addicted to the goal of material improvement and acquisition that even when we have reached a comfortable lifestyle, we want more and more. It appears that humanity cannot help itself indulging in ostentatious displays of meaningless wealth. Humanity cannot seem to curb its most basic avaricious instincts and the powers of the multinational firms and advertisers to exploit them, further weakening any innate resolve most people have to live sustainable lifestyles.

We continue to make items that we do not need nor are essential to our lives. We continue to industrially produce in huge quantities such artefacts where the extraction of materials for their fabrication causes huge environmental harm. We then discard these items, often when they are still functional or only in minor disrepair. Our processes of extraction, manufacture and discard treat the Earth as a bottomless pit, an environmental sink and a waste-bin for our discards, as if it had limitless powers of assimilation and sequestering with impunity. We display this behaviour despite our understanding of the fact that so much of what we discard is immensely harmful to natural ecosystems, whether the pollutants be gases discharged into the atmosphere or non-biodegradable artefacts such as plastics discharged into surface waters.

It is clear to all of humanity except the most self-deluding that our activities as a species have now brought about extensive global environmental consequences exceeding many of Nature's thresholds of resilience. We have actually managed to modify global biogeochemical cycles. The crucial evidence here is anthropogenic climate change. There are some who still deny this, looking for gaps in the data and alternative explanations, but climatic models have been tested by several supercomputer systems against the effects of other (non-human) factors, and our signature remains clear. Predictions vary, but it is quite possible that we only have a matter of decades before the systemic collapse of many global ecosystems, when these will reach a point that will only be reversible over periods that are longer than the history of human civilisation, or indeed human existence. One example is the global acceleration in loss of coral reefs through bleaching as a result of anthropogenic changes in climate and ocean chemistry.

Of greatest concern is perhaps the spectacle of trying to revert to a view of the Planet where gross domestic product is equated with well-being and human progress, where countries can act as if they are independent from each other and as if the consequences of their actions can somehow remain largely within national borders. This endeavour shows a self-defeating lack of ecological literacy.

The fact is that human society is now no longer in a preventive mode, but engaged in a 'race and rescue' mission to address humanity's 'sustainability equation' and to 'save the Planet' if we are to secure as sustainable future for ourselves and other life on Earth. The 'business as usual' option is clearly and emphatically not one that is compatible with the long-term survival of human civilisation, and possibly not with the existence of human beings as a species. A new approach and ideology towards Nature is needed that places the science of ecology in a leading role at all stages in designing, making, operating and recovering artefacts and systems of our human-made world based on the principles of **ecocentric** and **ecomimetic** design.

Many others have elaborated on this condition of the Planet at greater length and depth than is the intention or mission to attempt in this book. However, far fewer have combined a holistic view of anthropogenic Planetary degradation with an ecocentric and ecomimetic holistic approach to design and environmental integration – that is to say, what we must now *do* about the current condition we have created and are continuing to exacerbate.

Human society is now primarily an urban species, and it is predicted that by the year 2100 more than 80% of the global human population may reside in urban centres. It is for this reason that the greatly inefficient and harmful designs of nearly all of our existing cities and urban conurbations need to be remade and those of our new cities and urban conurbations need to be rethought and reinvented.

The reinvention is not to seek at the onset immediate innovative solutions to the problem, but starts with identifying the causes of the problem and framing the problem in the context that enables innovation. The causes are self-evidently defined as the result of humanity's callous abuse of Nature. The problem is summarily framed as an overarching '**sustainability equation**' that consists of addressing simultaneously the following key goals:

- Providing human society and all other lifeforms with the basic and essential needs for living.
- Rectifying past and current anthropogenic environmental impairments.
- Addressing society's humanitarian issues.
- Remaking the human-made world and its urban built environment to be ecologically effective.

The importance of this is that so many approaches to sustainability today are incomplete. Even practising environmentalists tend to implement one aspect and ignore the others. Actions to 'solve' this equation must not be incremental, singular or piecemeal. They have to be concerted, collective and large in scale. They need to be at least at the infrastructural, and where possible at the bioregional and global, scale to be effective. The reasons for this should be clear from the foregoing – because of our current role as global ecological designers and also because of the ethical duty of stewardship of Nature that comes with this role. These goals for resolution of the sustainability equation translate into the need to implement a number of imperative directives for human society.

RESOLVING THE SUSTAINABILITY EQUATION

In Chapter 1, the term 'sustainability' was recast in the form of an overarching sustainability equation of simultaneous endeavours to meet the following key goals. To summarise, we must, by adopting an ecocentric and ecomimetic approach:

- Fully evaluate and account for the our potential ecological impacts on Nature, both positive and negative, covering all of our human-made activities and processes and all of the materials we produce throughout their life-cycles, clearly identifying trajectories and potential consequences of all our actions before we act.
- Phase out existing activities that are harming Nature, most especially those relating to environmental pollution, climate change and habitat destruction.
- Avoid causing further deleterious and irreversible harm to Nature.
- Take direct and immediate action to repair, restore, stabilise and, where possible, rejuvenate existing damaged ecological and biospheric systems to enable these to thrive. Even if it is assumed that the effects of restorative action are linear, by the time restorative action is taken, the consequences of humanity's devastative action will have impacted a large portion of the Planet's ecological life support base.

Addressing the environmental and human dilemmas summarised in our **sustainability equation** in Chapter 1 needs a clear focus on the multiple tasks stated in that chapter to repurpose, reinvent, redesign, remake and recover our human-made world so that our built environment is seamlessly biointegrated with Nature and functions synergistically with it, in the urban realm where so many of our resources are used and wastes produced.

Of course, much has been written on green design, green urbanism, green cities, even cities as 'sustainable ecosystems', but most reports fall short of the mark of expressing the degree and depth of holistic and systemic change that is required.

Moreover, many designers continue to design with their primary driver being their own personal (and frequently merely aesthetic and intellectual) design emphases that are not ecologically benign. If all architects, urban planners, engineers, landscape architects, and indeed ecologists, adopted the ecocentric and ecomimicry approach and principles set out here, then we would be living in a very different human-made world, but ostensibly this is not currently the case.

It is contended here that the we need to address all the dysfunctional aspects of human society's relationship to Nature and the need for the effective Ecological Design and biointegration of our various 'infrastructures'. Where critical, they must remade entirely and at a significantly effective scale to avoid further damage to Nature and to start restoring it, and to go even further, to rejuvenate it. It is contended that we do not have much time to do this, perhaps less than a human generation.

We can now summarise how this might be achieved.

THE ECOCENTRIC APPROACH

Crucial to all our endeavours to save the Planet is the ecocentric, ecology-based approach for our acts, activities and our built artefacts and systems on the Planet. Ecological Design is a comprehensive and rigorous undertaking – and as with all matters in dealing with the complexity of natural systems, is a complex endeavour. The legion of interactive pathways in and between ecosystems are intricate and varied. One result of this is that changes to and manipulation of one fragment causes a series of changes that reverberates throughout ecosystems. It is this complexity of linkages and potential impacts that calls for an ecocentric rather than purely technocentric approach to solving our current human-made environmental crisis.

Ecological Design therefore demands that we take into account the ecosystems, multiple linkages and multiple levels of effects that are interdependent. Achieving this requires consideration of the impacts on all lifeforms, habitats and processes that are affected by any designed artefact or system.

Ecological Design entails more than ensuring that the existence and location of any constructed system or artefact does not harm Nature in a given time and place of intervention. All components of any designed system or artefact need to be considered in depth throughout their life-cycles, along with the quantity and source of energy required to make each transition or change, to ensure ecological effectiveness.

As explained in Chapter 7, this means undertaking an ecology-based 'environmental impact assessment' of every life-cycle stage. These evaluations enable an informed basis for the selection of materials in the designed system and the way they are assembled and used. Designing, then, involves prioritising and trading off the materials to be used, and identifying and addressing selectively those crucial to the design programme and to the particular conditions pertaining to the given programme, its locality and its circumstances with the least environmental impacts.

The critical but not the sole metric of the effectiveness of Ecological Design is the extent that it is able not only to avoid and eliminate negative irreversible impacts on Nature, but to repair and enhance it, supporting the integrity of natural and semi-natural ecosystems.

Accordingly, the key end objective of Ecological Design is to become 'at one with Nature'. This might sound trite or emotive, but is a vital goal. It means that humans achieve, in contrast to their current state of near 'war' with Nature, a symbiotic state of seamless and benign physical and systemic biointegration with it. In other words, we have to align the construction and operation of our human-made world – and most especially the built environment – with the needs and operations of natural and semi-natural ecosystems and the biogeochemical cycles of Planet Earth.

A key means in our endeavour to save and to reclaim Nature is through design. This is a problem-solving approach based on prototyping and testing. But this must in all ways and absolutely be guided by the science of ecology if we are to achieve a sustainable, durable and resilient future for all lifeforms and their habitats on

Earth. This means emulating and replicating ecosystem attributes in technological systems, making our built environment share as many of the attributes of a fully functional natural ecosystem as possible and bringing the two together in an ecologically effective 'biointegration'.

There needs to be *physical biointegration* – the spatial placement, arrangement, configuration and modification of the built environment and all of its synthetic human-made artefacts with ecosystems and habitats to avoid net harm to Nature and, where possible, enhance it (by creating new opportunities for biodiversity). How we can do this is summarised further later in this chapter.

There also needs to be *systemic biointegration* – the alignment of the built environment's systems, operations, processes and flows, including its inputs and outputs or materials and energy, with those of natural and semi-natural ecosystems and the Earth's biogeochemical cycles in a dynamic but generally stable equilibrium. There also needs to be a systemic biointegration of our urban structures and systems with our human social (anthropocentric) systems.

Achieving both physical and systemic biointegration is *ecologically effective biointegration*, and this means achieving a condition where the built environment and Nature are linked in a seamless ecological nexus. This is what is meant by the chapter title, 'Being 'at one' with Nature'.

To get to this point, we need, through the study of ecosystems and mimicking their characteristics, to reinvent, redesign, remake, recover and reconnect our human-made world as far as we possibly can.

This is the ambitious and audacious goal of Ecological Design.

THE ECOMIMETIC APPROACH

Following from the ecocentric approach is designing based on 'ecomimicry' in an endeavour to ensure that the built environment adopts as far as possible the normative attributes of the 'ecosystem' so that it can provide the many functions of ecosystems as fully as possible. This term 'ecomimicry' is adapted from the term 'biomimicry', which focuses on technological infrastructure and basing its design on structures and systems seen in Nature. Ecomimicry as a concept expands and advances biomimicry to apply to the *holistic biointegration* of the technological, natural and social (see Chapter 4), both physically and systemically. In applying ecomimicry, the design process changes towards the emulation and replication of the normative and emergent ecosystem attributes, generalised characteristics and identities of the ecosystem abstracted from systems ecology. 'Emergent attributes of ecosystems' in this context refers not to a property or attribute of any component of the ecosystem, but features of the system as a whole.

To summarise, these are attributes we need to ensure that our created systems and artefacts:

Are fuelled by sustainable energy sources as close to direct fuelling by the sun's energy as possible.

Recycle all components, biotic and abiotic, including air and water, so efficiently that the concept of waste becomes all but obsolete.

Accommodate processes of ecological succession and change over time so that we create more management plans than masterplans.

Display robust properties of homeostasis so that they self-rectify in the face of environmental damage.

Display benign and synergistic interconnection one with the other and with natural and semi-natural systems.

Are sufficiently complex to display robust ecosystem integrity – that is, they are able to show substitution in the event of significant harm to or loss of certain components so that other components take on the role of the component lost or harmed.

Besides these, there are other ecosystem properties. But most importantly, all of these attributes must be emulated and replicated to achieve the greatest **multifunctionality** and efficiency in delivery of meaningful **ecosystem services**. The nature and types of ecosystem services were elaborated in Chapter 5. Ecosystems provide these services to humanity without being asked, and the costs of having to replace these functions with human labour and technology have only become clear and widely reported in recent decades. As we seem so adept at destroying Nature's capacity to support us, we need to become equally adept at replacing this capacity. But in so doing, we must work with Nature and apply all our creative energies to new synergistic designs that maximise ecosystem service provision as never before.

While in the current world of design each of these attributes and functionalities has been emulated to one extent or another, ecomimesis requires their collective and holistic emulation to create complete and self-contained constructed ecosystems. The effective emulating and replicating of *all* of these attributes and properties are *key tenets* of Ecological Design that must pervasively guide all our human action in the making and remaking of our built systems. We need to accomplish this in totality within the limits of resilience of Nature's ecosystems and global cycles – the crucial limiting factors and parameters of Ecological Design.

THE HOLISTIC BIOINTEGRATION OF INFRASTRUCTURES AND MAKING BUILT ENVIRONMENTS AS 'QUATROBRID' CONSTRUCTED ECOSYSTEMS

It has been argued that the proximate aim of ecomimesis is to ensure that we design to provide for and/or enhance the provision of ecosystem services by Nature – the 'life support system' not just for humankind, but for all lifeforms. The proposal here is that this can be achieved by *working with* Nature and integrating it seamlessly into a new union with our technological and socio-political structures and systems, thereby creating new, multifunctional, ecologically effective systems.

Redesigning and reinventing the built environment and other human artefacts according to the principles of ecomimesis has been practically framed as the holistic integration of four 'infrastructures' (both natural and human-made). These partly overlap, but can be distinguished for ease of practical application:

Nature-based infrastructure (all the biotic components and processes based on species other than humankind).

Hydrological infrastructure (water as part of Nature is singled out due to its fundamental importance to known life).

Technological infrastructure (everything we make and operate) constructed ecosystems.

Anthropocentric infrastructure (our socio-political and cultural institutions and systems), which is mainly abstract in nature.

Combining all the attributes of Nature-based, hydrological, technological and anthropocentric infrastructures into a single analytical matrix facilitates the inclusive design of 'quatrobrid' ecologically augmented 'constructed ecosystems'. Such systems are envisaged as no longer independent of natural and semi-natural systems and processes, but acting as new, synergistic ecosystems that help to positively and proactively support and restore the integrity of wider Nature within the framework of new dynamic equilibria.

The emergent product of this endeavour is part biotic, part technological and cybernetic, and also socially and culturally responsive and proactive. In this regard, the vision has much in common with the 'eco-cyborg' that is referred in the world of science fiction. It is a system that can be optimised and which is adaptive, changing the environmental conditions.

The need to *restore* Nature is acute. We can no longer work and design simply to minimise our adverse environmental impacts. We need to develop a human-made world in ways that results in *net enhancement* of Nature, not least because in order to have greater opportunities and resilience for the future, we need to repair and make good past harm we have caused to Nature. Hence, *every design project* becomes one of environmental repair, restoration, enhancement and augmentation.

The tasks, then, of reinventing and recreating our built environment and all of the artefacts we make is no longer an anticipatory or preventive undertaking of 'designing today for tomorrow'. It has as its core elements the urgent endeavours of restoring and rejuvenating Nature.

In terms of urban design, our quatrobrid constructed ecosystems can take the physical form of anything from a specific localised intervention or artefact to whole integrated biotic/vegetated systems weaving through every element of the urban fabric. They can, and must, take the form of systemically integrating the processes and flows of the biotic and abiotic constituents of the built environment with those of natural and semi-natural ecosystems. This is seen as being achieved without any displacement of valued Nature in the interactions. It is seen as maintaining the integrity of ecosystems in terms of the web of species, functions and processes crucial for ensuring stability and resilience.

All design endeavours involve prototyping, and in ecological, ecomimetic design, the prototyping of the 'constructed ecosystem' is absolutely vital. Being part biotic, such systems will inevitably evolve, and indeed mutate. It will take time to determine the correct system components with the correct systems of management and intervention, for example, to arrest successional processes at the point of optimum delivery of ecosystem services.

Achieving this integration to make a new ecological urbanism a reality requires the creation of a global, extensive, complex, dynamic and networked system of monitoring and instantaneous response. This is clearly a major undertaking, but one that is vital to make an impact on a scale sufficient to address the current environmental crisis. It requires the real and full realisation of human society's much-vaunted 'Fourth Industrial Revolution' in which boundaries between the biotic, abiotic and cybernetic are completely blurred.

There will always be limitations and inadequacies, both in terms of our understanding of ecosystem function and in our development and applications of technology. However, we cannot permit such issues to inhibit the urgent implementation of ecomimesis for the creation and transformation of the existent built environment and all associated production systems and artefacts. Humanity's endeavours to address environmental decline need to be an ongoing developmental process, but have to be immediately accelerated.

HUMAN IDEOLOGIES

An excursion was made in Chapter 8 into the world of human socio-economic-political and cultural systems that form, in design terms, our 'anthropocentric' infrastructure. The key message here is that our future and that of so many other species depends on whether or not our ideological framework can become universally ecocentric. All of human civilisation depends, certainly for the readily foreseeable future, fundamentally on the continued ecological integrity of the Planet. Being a closed system is referred to by some as 'spaceship Earth'.

In terms of human ideology and habits, humanity as a species needs to conquer the addiction to self-serving materiality that causes net environmental harm and fuels a measure of growth that omits the most important consideration: human happiness and health. We need to rebut the traditional dichotomies between human prosperity and the environment by establishing clear new definitions of what true prosperity and richness of life actually means. Designers can play a significant role in this by showing what is possible, what economic and other intangible benefits can accrue and what new human endeavours can be supported by such an approach. However, it is for the thought leaders and policy makers to take up the mission from the very top. We need our leaders and governments to fully *facilitate and incentivise rather than oppose* any grassroots societal movements towards ecotopia – a new vision of human civilisation based on environmental integrity, respect for all lifeforms, human social equity and human happiness.

HOW DO THE IDEAS, PRINCIPLES AND DIRECTIVES PRESENTED HERE IMPACT OUR HUMAN SOCIETY?

What are the significances and impacts of the ideas, principles and directives that are set out here?

The two most important ideas are **ecocentricity** (action guided by and based on the science of ecology) and **ecomimesis** (action based on the emulation and replication of the attributes of the 'ecosystem' concept). These together radically change the way we think about Nature (our ideology and our role and relationship with Nature), the way we live and work (in relation to and symbiosis with the ecosystems and the biogeochemical cycles in Nature) and the way we design, make, manage, operate and recover our existing and new built environment, as the biointegration of sets of infrastructures into quatrobrid constructed ecosystems. For the designer, it changes forever the way that the designing of our built environment and all our artefacts is carried out.

If we are to further develop these ideas, principles and directives, these in totality can be regarded broadly as a 'manifesto' that needs to be further developed in culturally, socially, economically and politically specific ways where every race and nation on Earth can implement them. Provided here are the initial and actionable premises and bases for humanity to repair and to achieve a harmonious relationship between our human-made world and the natural systems in the Planet by:

Adopting the science of ecology as the fundamental discipline that must guide all of humanity's endeavours on Earth.

Working within Nature's thresholds of resilience and its limits of provision of non-renewable natural resources.

Redesigning and remaking the human-made world based on the principles of ecomimicry, emulating and replicating the normative attributes of ecosystems in the built environment and maximising delivery of ecosystem services while supporting the integrity of natural and semi-natural ecosystems. Its production systems and its artefacts become hybrid 'human-made ecosystems'.

Biointegrating all essential components of true Ecological Design as composite durable, sustainable and holistic quatrobrid 'constructed ecosystems' combining natural, hydrological, technological and anthropocentric infrastructures.

Re-envisioning our human-made world, and in particular our built environment, as 'ecotopias' that incorporate all of the above to achieve a sustainable Planet in which human civilisation can survive alongside the rest of biodiversity. We need to apply this vision to both our new and our existing urban conurbations, weaving, fusing and biointegrating multifunctional Nature-based and hydrological infrastructures throughout.

Inherent in the concept of 'sustainability' is concern for future generations, where sustainability requires *'development that meets the needs of the present without compromising the ability of future generations to meet their own needs'* and promotes *'harmony among human beings and between humanity and nature'.*[1] The questions posed now are how many future generations will there be, what are their needs, and what kind of harmony do they hope to achieve? It is evidently clear, then, that our society has now reached a point in our biological, social and cultural evolutions as a species where, due to our massive ability to impact significantly and negatively Nature's processes on the Planet, the number of our future generations that we can safeguard without urgent action may well be pitifully few.

Our human society has a duty of total stewardship over the Planet Earth in seeking harmony with Nature. A practical, holistic and integrated approach needs to be adopted by all of humanity directly in the course that is defined here as **ecocentricity** and **ecomimesis**, brought together by effective **biointegration**. These, together with the other directives and principles described here, must be applied globally, imminently, collectively and in synchrony by all of humanity, as all of human society is affected.

NOTE

1 World Commission on Environment and Development. *Our Common Future*. Oxford University Press, 1987, pp. 43 and 65.

GLOSSARY

Abiotic Factors inorganic constituents, or the physical features of an ecosystem.

Anthropocene the current geological era in which human activity is the dominating influence over Earth's systems.

Anthropocentric the belief that human beings are the most significant species in the universe, operating with a clear priority for the interests of humans with little regard for the merit of other species.

Anthropocentric Infrastructure infrastructure that is completely related to the existence and interests of humankind. The combination of humanity's systems, organisations, rules, ideologies and ideas can be thought of as a part-physical, part-virtual infrastructure.

Anthropogenic caused or influenced by humans.

Applied Ecology the application of the science of ecology integrated with social and biotechnical aspects to address natural resource management and conservation.

Artificial Intelligence the theory and development of computer systems able to perform tasks that normally require human intelligence, such as visual perception, speech recognition, decision-making and translation between languages.

Assimilation Capacity the capacity of an ecosystem or ecosystem component (such as a body of water) to receive waste or toxic substances without damaging the environment or harming those who consume the water.

Atmosphere the layer of gases surrounding the Earth, keeping moisture and elements in while protecting the Planet from harmful ultraviolet radiation and extreme temperatures.

Atmospheric Cycles the cycling of elements and compounds through physical processes in and out of the atmosphere.

Biodiversity the variability that exists within species, between species and between ecosystems, as well as the rate of change of that variability across time or space.

Biogeochemical Cycles the cycling of elements and compounds through biological, geological, and chemical processes in Earth's systems. This includes the water, carbon and nitrogen cycles, among others.

Biological Interaction the effects a pair of organisms living together in a community have on each other. They can be either of the same species or of different species.

Biomimicry the emulation of biological organisms and processes in the design of structures and materials.

Biophilia human affinity to Nature and natural designs, as well as an innate tendency to seek connection to Nature.

Bioremediation the use of biological agents, such as bacteria, to clean up toxic or hazardous contamination created by human-caused pollution.

Biosphere the global ecological system, encompassing all living things and their interactions

Biotechnology the exploitation of biological processes for industrial and other purposes, especially the genetic manipulation of microorganisms for the production of antibiotics, hormones etc.

Biotic Constituents biological organisms, or the interacting species in an ecosystem.

Blackwater sewage, or wastewater from toilets that cannot be reused unless properly treated.

Blue Infrastructure a network providing the 'ingredients' for solving urban and climatic challenges by building with Nature. Blue infrastructure specifically relates to landscape elements linked to water (pools, lakes, ponds, watercourses etc.)

Carbon Sequestration the ability of abiotic and biotic components of an ecosystem to uptake carbon and serve as a 'carbon sink', keeping CO_2 out of the atmosphere.

Carrying Capacity of a species: the maximum population an ecosystem can sustain given the available resources.

Circular Economy a circular economy aims to redefine growth, focusing on positive society-wide benefits. It entails gradually decoupling economic activity from the consumption of finite resources and designing waste out of the system. Underpinned by a transition to renewable energy sources, the circular model builds economic, natural and social capital. It is based on three principles: (1) design out waste and pollution; (2) keep products and materials in use; (3) regenerate natural systems (Ellen MacArthur Foundation).

Climax Community a group or association of two or more populations of different species that achieve a steady state, finding a balance between their interactions and coexisting.

Closed System a system in which all outputs are inputs for another component in the system and nothing is considered 'waste'.

Combined Heat and Power Plant also known as cogeneration, a suite of technologies that can use a variety of fuels to generate electricity or power at the point of use, allowing the heat that would normally be lost in the power generation process to be recovered to provide needed heating and/or cooling.

Commensalism an association between two organisms in which one benefits and the other derives neither benefit nor harm.

Community a group of two or more populations of different species occupying the same place and time.

Consumers a class of living organisms that get energy by consuming other organisms rather than creating energy themselves by converting abiotic components.

Critical Ecological Threshold the point at which a relatively small change or disturbance causes a rapid change in an ecosystem.

Cultural Eutrophication the process that speeds up natural eutrophication because of human activity. Due to clearing of land and building of towns and cities, land runoff is accelerated and more nutrients such as phosphates and nitrates are supplied to lakes and rivers, and then to coastal estuaries and bays. These nutrients cause an explosion in algal populations, or harmful algal blooms, which deplete the water of oxygen and block sunlight, which is deadly for many species in aquatic ecosystems.

Decomposers an organism, especially a soil bacterium, fungus or invertebrate, that decomposes organic material.

Diurnal anything that occurs during the daytime. It can also refer to species only active during the day.

Ecocentricity applying the science of ecology first and foremost in the design process.

Ecological Fragmentation when habitat is separated by natural or human-made physical barriers such as telephone wires, highways or suburban residential developments. This can be highly detrimental for migrating species such as birds. Restoration efforts include creating 'wildlife corridors' by protecting connecting land for migrating species to travel through.

Ecological Impact Assessment an environmental impact assessment is required in the US for many construction and infrastructure projects to evaluate likely environmental (and ecological) impacts caused by a development project, including socio-economic, cultural and human health impacts.

Ecological Integrity an ecosystem's ability to maintain processes to support a diverse community of biological organisms and species.

Ecological Reinventing the process of creating new and reconstituted human-made systems that function benignly in Nature.

Ecological Resilience the capacity of an ecosystem to recover from an external disturbance.

Ecological Succession the process by which biological communities take over a region and evolve over time. Primary succession occurs in essentially barren or lifeless areas, such as in the wake of a lava flow or retreating glacier, while secondary succession occurs after environmental disturbance, such as a forest fire or extreme weather event.

Ecologically Effective the ability of an urban centre to interact benignly with the ecology of the Planet's systems.

Ecology the study of how biotic and abiotic constituents in an ecosystem interact.

Ecomimicry the emulation of ecological systems in the design of the built environment.

Ecosystem the biotic constituents of a community and the abiotic factors of their physical environment.

Ecosystem Health this is a subjective term often up for debate, but definitions tend to include reference to stability, being sustainably able to maintain autonomy over time, and resiliency.

Ecosystem Productivity the production of biomass of an ecosystem or the conversion of inorganic components into energy consumed and used by biotic components of an ecosystem.

Ecosystem Services the beneficial services produced by a healthy, fully functioning ecosystem, including water and air purification, waste management, pest control, climate regulation and nutrient cycling.

Embodied Energy the sum of all the energy required to produce any goods or services, considered as if that energy was incorporated or 'embodied' in the product itself – the total of the energy necessary for a designed object or system's entire life-cycle through raw material extraction, transport, manufacture, assembly, installation, disassembly, deconstruction and/or decomposition.

Environmental Biology the study of specific regional environments as well as the organisms and wildlife that inhabit those locations. Environmental biologists aid in the protection and survival of wildlife within any given ecosystem, and assess the impact of human activity on the environment.

Environmental Sink in this system of biogeochemical cycles where every output is another's input, a 'source' releases a chemical element or compound, while a 'sink' receives that element or compound. Biogeochemical processes occur on different timescales, so a sink often refers to a component that holds an element or compound for a relatively long time, such as carbon dissolved in the ocean.

Evolution the natural selection of advantageous heritable traits in successive generations.

Extremophile a microorganism, especially an archaean, that lives in conditions of extreme temperature, acidity, alkalinity or chemical concentration.

Geosphere the solid portion of the Earth from the crust towards the centre, including the core and the inner mantle layer.

Global Climate Change 'a change in the state of the climate that can be identified (e.g. using statistical tests) by changes in the mean and/or the variability of its properties, and that persists for an extended period, typically decades or longer' (Intergovernmental Panel on Climate Change, 2007). 'Most climate scientists agree the main cause of the current global warming trend is human expansion of the "greenhouse effect" – warming that results when the atmosphere traps heat radiating from Earth toward space' (NASA, 2017).

Green Infrastructure a network providing the 'ingredients' for solving urban and climatic challenges by building with Nature. Green infrastructure specifically relates to landscape elements linked to vegetation (parks, trees, gardens etc.).

Greywater relatively clean wastewater from baths, sinks, washing machines and other kitchen appliances that can be used for irrigation or other second uses.

Habitat Fragmentation the process by which habitat loss results in the division of large, continuous habitats into smaller, more isolated remnants. Habitat fragmentation describes the emergence of discontinuities in an organism's preferred environment, causing population fragmentation and ecosystem decay.

Hexavalent Chromium hexavalent chromium is any chemical compound that contains the element chromium in the +6 oxidation state, Cr(VI). Virtually all chromium ore is processed via hexavalent chromium, specifically the salt sodium dichromate. Approximately 136,000 tonnes of hexavalent chromium were produced in 1985. Cr(VI) compounds may be used as pigments in dyes, paints, inks and plastics. It may also be used as an anticorrosive agent added to paints, primers and other surface coatings. The Cr(VI) compound chromic acid is used to electroplate chromium onto metal parts to provide a decorative or protective coating. Irritation or damage to the eyes and skin can occur if hexavalent chromium contacts these organs in high concentrations. Breathing in high levels of hexavalent chromium can cause irritation to the nose and throat. Repeated or prolonged exposure can cause sores to develop in the nose and result in nosebleeds.

Holistic Urban Design rather than focusing on one aspect or targeted design feature, considering all relevant systems as well as how they will interact with each other and with natural Planetary cycles.

Homeostasis an ecosystem's ability to self-regulate and find balance after external disturbances or perturbations.

Hydrology 'the science that treats the waters of the Earth, their occurrence, circulation, and distribution, their chemical and physical properties, and their reaction with the environment, including the relation to living things' (National Research Council, 1991).

Hydrosphere all of the water on Earth's surface, including water in the atmosphere.

Internet of Things the interconnection via the internet of computing devices embedded in everyday objects, enabling them to send and receive data.

Invasive Species a species introduced into a new ecosystem (through human or non-human vectors) that is able to vastly out-compete and destroy native species to a degree that harms ecosystem, economic and human health.

Keystone Species a species that has a large impact on the ecosystem disproportionate to its population size. Such a species plays a significant role in maintaining homeostasis and often provides major ecosystem services. Without this species, the ecosystem will take on entirely different characteristics and be more vulnerable to disturbance and invasion.

Life-Cycle Assessment considering the whole life-cycle of a material or product from the mining or harvesting of materials used to synthesise that material to the end use, including all emissions and effluents.

Linear Economy raw materials are used to make a product, and after its use any waste (e.g. packaging) is thrown away. In an economy based on recycling, materials are reused.

Mutualism interactions between organisms of two different species, in which each organism benefits from the interaction in some way.

New Communities as a biological system proceeds through the process of ecological succession, new communities of organisms emerge.

Nanotechnology the branch of technology that deals with dimensions and tolerances of less than 100 nanometres, especially the manipulation of individual atoms and molecules.

Parasitism a relationship between two organisms where one benefits and the other is harmed. The parasite is the organism that benefits from the relationship, while the host is harmed by the relationship. Parasites can include plants, animals, and even viruses and bacteria.

Peak Supply a supply of energy which will be used to augment existing energy sources during periods of peak demand. As examples, the noontime output of a hydroelectric generator is not considered to be a peak supply.

Peaked Hydrograph a hydrograph is a graph showing the rate of flow (discharge) versus time past a specific initial point. Peak discharge is the highest point on the hydrograph, where the rate of discharge is greatest.

Population Equivalent (or unit per capita loading) in wastewater treatment is the number expressing the ratio of the sum of the pollution load produced during 24 hours by industrial facilities and services to the individual pollution load in household sewage produced by one person in the same time.

Primary Producers an autotrophic organism producing complex organic matter using photosynthesis or chemosynthesis. Such an organism produces its own energy from abiotic constituents of an ecosystem rather than by consuming other organisms.

Quantum Computing the area of study focused on developing computer technology based on the principles of quantum theory, which explains the nature and behaviour of energy and matter at the quantum (atomic and subatomic) level.

Reference Point/Shifting Baselines a reference point is a baseline for measuring ecosystem health; however, that baseline is shifting over time as generations lose knowledge and perspective and societal memory are replaced.

Regenerative Action action humans can take to act as stewards of the Planet and work to restore ecosystem health.

Renewable Energy energy collected or produced from renewable resources, which are naturally replenished within a human timescale, such as sunlight, wind, rain, tides, waves and geothermal heat.

Resilience the ability of an ecosystem to maintain its normal patterns of nutrient cycling and biomass production after being subjected to damage caused by an ecological disturbance.

Semi-Natural Ecosystem an ecosystem which has been altered by human actions, but which retains significant native elements.

Stratospheric Ozone Layer a region of the Earth's stratosphere that absorbs most of the sun's ultraviolet radiation. It contains high concentrations of ozone in relation to other parts of the atmosphere, although still small in relation to other gases in the stratosphere.

Sustainability 'the long-term ability to maintain an ecosystem or human society' (Ecology Global Network, 2011).

Sustainable Development 'development that meets the needs of the present without compromising the ability of future generations to meet their own needs' (*Brundtland Report*, 1987).

Symbiosis a mutually beneficial relationship between two or more species in which all parties benefit from interaction with each other.

Synanthropic species of flora or fauna of various kinds that live near, and benefit from, an association with humans and the somewhat artificial habitats humans create around them.

Tectonic Shifts the movement of the tectonic plates that make up Earth's crust, caused by geologic activity from the Planet's core.

Terrestrial relating to land-based ecosystems, as opposed to aquatic freshwater or saltwater ecosystems.

Trophic Levels a level or hierarchy in the ecological food chain. A trophic pyramid represents the direction energy flows through the biotic constituents of an ecosystem, and each level represents all of the species that exist in that position of the energy flow. For example, primary producers are at the base of the trophic pyramid because they produce their own energy from inorganic constituents, either from sunlight or chemicals present in their surroundings.

Urban Avoiders organisms that are very sensitive to the changes in habitat that occur with urbanisation. They are the first species to disappear in urban settings. With constantly growing human population, urban sprawl continues to consume native landscapes.

Urban Dwellers species that can live or thrive in urban environments. Some urban wildlife, such as house mice, are synanthropic, ecologically associated with humans. Some species or populations may become entirely dependent on humans. Different types of urban areas support different kinds of wildlife.

Water Cycle the cycle of processes by which water circulates between the Earth's oceans, atmosphere and land, involving precipitation as rain and snow, drainage in streams and rivers, and return to the atmosphere by evaporation and transpiration.

Watershed a drainage basin or catchment area for a particular water body. The area of land that feeds water into that water body.

Wildlife Corridor a strip of natural habitat connecting populations of wildlife otherwise separated by cultivated land, roads etc.

BIBLIOGRAPHY

Andersson, E., et al. "Reconnecting Cities to the Biosphere: Stewardship of Green Infrastructure and Urban Ecosystem Services." *Ambio*, vol. 43, no. 4, 17 Apr. 2014, pp. 445–453. doi:10.1007/s13280-014-0506-y.

Baldé, C. P., V. Forti, V. Gray, R. Kuehr, and P. Stegmann. *The Global E-Waste Monitor 2017*. United Nations University, International Telecommunication Union and International Solid Waste Association, 2017.

Bao, K., and J. Drew. "Traditional Ecological Knowledge, Shifting Baselines, and Conservation of Fijian Molluscs." *Pacific Conservation Biology*, vol. 23, no. 1, 2017, p. 81. doi:10.1071/pc16016.

Benyus, J. M. *Biomimicry: Innovation Inspired by Nature*. Harper Perennial, 2009.

Berner, R. A. *The Phanerozoic Carbon Cycle: CO_2 and O_2*. Oxford University Press, 2004.

Bertalanffy, L. V. *General System Theory: Foundations, Development, Applications*. Braziller, 1980.

Bunker, A. F. "Computations of Surface Energy Flux and Annual Air–Sea Interaction Cycles of the North Atlantic Ocean." *Monthly Weather Review*, vol. 104, no. 9, 1976, pp. 1122–1140. doi:10.1175/1520-0493(1976)1042.0.co;2.

Cain, M. L., et al. *Ecology*. Sinauer Associates, Inc. Publishers, 2014.

Chang, K. "Hydrogen Cars Join Electric Models in Showrooms." *New York Times*, 18 Nov. 2014.

Chappell, J. "Coral Morphology, Diversity and Reef Growth." *Nature*, vol. 286, no. 5770, 1980, pp. 249–252. doi:10.1038/286249a0.

City of New York, Office of the Mayor. *NYC Green Infrastructure Plan: A Sustainable Strategy for Clean Waterways*. New York City Department of Environmental Protection, 2010.

Collias, N. E. and E. C. Collias. *Nest Building and Bird Behavior*, vol. 857. Princeton University Press, 2016.

Collinge, S. K. "Ecological Consequences of Habitat Fragmentation: Implications for Landscape Architecture and Planning." *Landscape and Urban Planning*, vol. 36, no. 1, 11 July 1996, pp. 59–77. doi:10.1016/s0169-2046(96)00341-6.

Costanza, R., et al. *Ecosystem Health: New Goals for Environmental Management*. Island Press, 1992.

Dang, T. D., et al. "Hydrological Alterations from Water Infrastructure Development in the Mekong Floodplains." *Hydrological Processes*, vol. 30, no. 21, 29 Apr 2016, pp. 3824–3838. doi:10.1002/hyp.10894.

Darwin, C. *On the Origin of Species*. John Murray, 1859.

Despommier, D. *The Vertical Farm: Feeding Ourselves and the World in the 21st Century*. Thomas Dunne Books, 2010.

Didham, R. K. "Ecological Consequences of Habitat Fragmentation." In *Encyclopedia of Lifc Sciences*. John Wiley & Sons, 2010, doi:10.1002/9780470015902.a0021904.

Duarte, C. M., et al. "Return to *Neverland*: Shifting Baselines Affect Eutrophication Restoration Targets." *Estuaries and Coasts*, vol. 32, no. 1, 2009, pp. 29–36. doi:10.1007/s12237-008-9111-2.

Edwards, P. E. T., et al. "Investing in Nature: Restoring Coastal Habitat Blue Infrastructure and Green Job Creation." *Marine Policy*, vol. 38, 12 June 2012, pp. 65–71. doi:10.1016/j.marpol.2012.05.020.

Elimelech, M. "Overview on Nanotechnology: The Future of Seawater Desalination." *International Journal of Science and Research*, vol. 5, no. 2, 2016, pp. 399–401. doi:10.21275/v5i2.nov161131.

Environmental Literacy Council. "Biogeochemical Cycles." 2018, https://enviroliteracy.org/airclimate-weather/biogeochemical-cycles/.

Environmental Literacy Council. "Carbon Cycle." 2018, https://enviroliteracy.org/air-climate-weather/biogeochemical-cycles/carbon-cycle/.

Environmental Literacy Council. "Sources & Sinks." 2018, https://enviroliteracy.org/air-climate-weather/climate/sources-sinks/.

Eregno, F. E., et al. "Quantitative Microbial Risk Assessment Combined with Hydrodynamic Modelling to Estimate the Public Health Risk Associated with Bathing after Rainfall Events." *Science of the Total Environment*, vol. 548–549, 2016, pp. 270–279. doi:10.1016/j.scitotenv.2016.01.034.

Ernest, S. K. M., and J. H. Brown. "Homeostasis and Compensation: The Role of Species and Resources in Ecosystem Stability." *Ecology*, vol. 82, no. 8, 2001, pp. 2118–2132. doi:10.2307/2680220.

Fierer, N. "Embracing the Unknown: Disentangling the Complexities of the Soil Microbiome." *Nature Reviews Microbiology*, vol. 15, no. 10, 21 Aug. 2017, pp. 579–590. doi:10.1038/nrmicro.2017.87.

Finlay, J. C., et al. "Human Influences on Nitrogen Removal in Lakes." *Science*, vol. 342, no. 6155, 11 Oct. 2013, pp. 247–250. doi:10.1126/science.1242575.

Flynn, K. J., et al. "Ocean Acidification with (De)Eutrophication Will Alter Future Phytoplankton Growth and Succession." *Proceedings of the Royal Society B: Biological Sciences*, vol. 282, no. 1804, 2015, p. 20142604. doi:10.1098/rspb.2014.2604.

Ford, T. E., and R. J. Naiman. "Alteration of Carbon Cycling by Beaver: Methane Evasion Rates from Boreal Forest Streams and Rivers." *Canadian Journal of Zoology*, vol. 66, no. 2, 1988, pp. 529–533. doi:10.1139/z88-076.

Foster, J., et al. *The Value of Green Infrastructure for Urban Climate Adaptation*. Center on Clean Air Policy, 2011.

Funch, R. R. "Termite Mounds as Dominant Land Forms in Semiarid Northeastern Brazil." *Journal of Arid Environments*, vol. 122, 2015, pp. 27–29. doi:10.1016/j.jaridenv.2015.05.010.

Gough, C. M. "Terrestrial Primary Production: Fuel for Life." *Nature Education Knowledge*, vol. 3, no. 10, ser. 28, 2011, p. 28, www.nature.com/scitable/knowledge/library/terrestrial-primary-production-fuel-for-life-17567411.

Gunderson, L. H. "Ecological Resilience – in Theory and Application." *Annual Review of Ecology and Systematics*, vol. 31, no. 1, 2000, pp. 425–439. doi:10.1146/annurev.ecolsys.31.1.425.

Hansen, R., and S. Pauleit. "From Multifunctionality to Multiple Ecosystem Services? A Conceptual Framework for Multifunctionality in Green Infrastructure Planning for Urban Areas." *Ambio*, vol. 43, no. 4, 17 Apr 2014, pp. 516–529. doi:10.1007/s13280-014-0510-2.

Harris, J. "Soil Microbial Communities and Restoration Ecology: Facilitators or Followers?" *Science*, vol. 325, no. 5940, 2009, pp. 573–574. doi:10.1126/science.1172975.

Heymans, J. J., et al. "Global Patterns in Ecological Indicators of Marine Food Webs: A Modelling Approach." *PLOS ONE*, vol. 9, no. 4, 2014, doi:10.1371/journal.pone.0095845.

Intergovernmental Panel on Climate Change. *Special Report on Global Warming of 1.5 °C (SR15)*. Intergovernmental Panel on Climate Change, 2018.

Johnson, L. E. "GIS and Remote Sensing Applications in Modern Water Resources Engineering." In L. Wang and C. Yang (eds.), *Modern Water Resources Engineering: Handbook of Environmental Engineering*, vol. 15. Humana Press, 2014, pp. 373–410.

Jones, C. G., et al. "Organisms as Ecosystem Engineers." *Oikos*, vol. 69, no. 3, 1994, pp. 373–386. JSTOR: www.jstor.org/stable/3545850.

Kaminski, L. A., et al. "Interaction between Mutualisms: Ant-Tended Butterflies Exploit Enemy-Free Space Provided by Ant-Treehopper Associations." *The American Naturalist*, vol. 176, no. 3, 20 July 2010, pp. 322–334. doi:10.1086/655427.

Karamouz, M. *Hydrology and Hydroclimatology: Principles and Applications*. CRC Press, 2013.

Karl, T. R. "Global Climate Change Impacts in the United States." In T. R. Karl, J. M. Melillo, and T. C. Peterson (eds.), *Global Climate Change Impacts in the United States: A State of Knowledge Report from the U.S. Global Change Research Program*. Cambridge University Press, 2009, pp. 13–52.

Kazmierczak, A., and J. Carter. *Adaptation to Climate Change Using Green and Blue Infrastructure: A Database of Case Studies*. University of Manchester, June 2010.

Lal, R. "Carbon Sequestration." *Philosophical Transactions of the Royal Society B: Biological Sciences*, vol. 363, no. 1492, 30 Aug. 2007, pp. 815–830. doi:10.1098/rstb.2007.2185.

Liss, K. N., et al. "Variability in Ecosystem Service Measurement: A Pollination Service Case Study." *Frontiers in Ecology and the Environment*, vol. 11, no. 8, 26 June 2013, pp. 414–422. doi:10.1890/120189.

Lovins, A. B. "More Profit with Less Carbon." *Scientific American*, vol. 293, no. 3, 2005, pp. 74–83. doi:10.1038/scientificamerican0905-74.

Lu, Y., et al. "Ecosystem Health towards Sustainability." *Ecosystem Health and Sustainability*, vol. 1, no. 1, 2015, pp. 1–15. doi:10.1890/ehs14-0013.1.

Margulis, S. *Introduction to Hydrology*, 4th ed., 2017, https://margulis-group.github.io/teaching/.

Mathez, E. A. and J. E. Smerdon. *Climate Change: The Science of Global Warming and Our Energy Future*. Columbia University Press, 2018.

Molla, A. S. "Natural, Semi-natural, and Artificial Ecosystems." *The Daily Star*, 29 Nov. 2008.

More, T. *Utopia* [1516]. Translated by P. K. Marshall. Washington Square, 1965.

National Geographic. "Biosphere." *National Geographic Resource Library*, 9 Oct. 2012, www.nationalgeographic.org/encyclopedia/biosphere/.

Ocean Health Index. "Ecological Integrity." 2019, www.oceanhealthindex.org/methodology/components/ecological-integrity.

O'Connor, F. M., et al. "Possible Role of Wetlands, Permafrost, and Methane Hydrates in the Methane Cycle under Future Climate Change: A Review." *Reviews of Geophysics*, vol. 48, no. 4, 2010, doi:10.1029/2010rg000326.

Olson, R. "Slow Motion Disaster below the Waves." *LA Times*, Sunday Opinion Section, Environment, 17 Nov. 2002, www.shiftingbaselines.org/op_ed/.

Papworth, S. K., et al. "Evidence for Shifting Baseline Syndrome in Conservation." *Conservation Letters*, 2009, doi:10.1111/j.1755-263x.2009.00049.x.

Pawlyn, M. *Biomimicry in Architecture*. RIBA Publishing, 2011.

Peterson, G. D., C. R. Allen, and C. S. Holling, "Ecological Resilience, Biodiversity, and Scale." *Nebraska Cooperative Fish and Wildlife Research Unit – Staff Publications*, 4, 1998. https://digitalcommons.unl.edu/ncfwrustaff/4.

Pimm, S. L., and J. H. Lawton. "Number of Trophic Levels in Ecological Communities." *Nature*, vol. 268, no. 5618, 17 May 1977, pp. 329–331. doi:10.1038/268329a0.

Pugh, T. A. M., et al. "Role of Forest Regrowth in Global Carbon Sink Dynamics." *Proceedings of the National Academy of Sciences*, 19 Feb. 2019, doi:10.1073/pnas.1810512116.

Rapport, D. "Assessing Ecosystem Health." *Trends in Ecology & Evolution*, vol. 13, no. 10, 10 Oct 1998, pp. 397–402. doi:10.1016/s0169-5347(98)01449-9.

Reisner, M. *Cadillac Desert: The American West and Its Disappearing Water*. Penguin, 1993.

Rittmann, B. E., and P. L. McCarty. *Environmental Biotechnology: Principles and Applications*. McGraw-Hill, 2001.

Schiermeier, Q., et al. "Energy Alternatives: Electricity without Carbon." *Nature*, vol. 454, no. 7206, 2008, pp. 816–823. doi:10.1038/454816a.

Selge, S., et al. "Public and Professional Views on Invasive Non-Native Species – a Qualitative Social Scientific Investigation." *Biological Conservation*, vol. 144, no. 12, 2011, pp. 3089–3097. doi:10.1016/j.biocon.2011.09.014.

Simberloff, D. "Confronting Introduced Species: A Form of Xenophobia?" *Biological Invasions*, vol. 5, no. 3, 2003, pp. 179–192. doi:10.1023/a:1026164419010.

Singh, V. P., A. K. Mishra, H. Chowdhary, and C. P. Khedun. "Climate Change and Its Impact on Water Resources." In L. K. Wang and C. T. Yang (eds.), *Modern Water Resources Engineering*. Humana Press, 2014, pp. 525–569.

Smith, V. H. "Cultural Eutrophication of Inland, Estuarine, and Coastal Waters." *Successes, Limitations, and Frontiers in Ecosystem Science*, 1998, pp. 7–49. doi:10.1007/978-1-4612-1724-4_2.

Sorokin, Y. I. *Coral Reef Ecology*, vol. 102. Springer, 1995.

Stockholm Resilience Centre. "The Nine Planetary Boundaries." 2019, www.stockholmresilience.org/research/planetary-boundaries/planetary-boundaries/about-the-research/the-nine-planetary-boundaries.html.

Tzoulas, K., et al. "Promoting Ecosystem and Human Health in Urban Areas Using Green Infrastructure: A Literature Review." *Landscape and Urban Planning*, vol. 81, no. 3, 2007, pp. 167–178. doi:10.1016/j.landurbplan.2007.02.001.

Von Wong, B. "4100lbs of E-Waste Resurrected. Trillions to Go." *Von Wong Blog*, 19 Apr. 2018, https://blog.vonwong.com/dell/.

WildMadagascar.org. "Why Rainforest Soils Are Generally Poor for Agriculture." 2019, www.wildmadagascar.org/overview/rainforests2.html.

Willis, K. J. and J. C. McElwain. *The Evolution of Plants*, 2nd ed.. Oxford University Press, 2014.

Wirsen, C. O., et al. "Chemosynthetic Microbial Activity at Mid-Atlantic Ridge Hydrothermal Vent Sites." *Journal of Geophysical Research*, vol. 98, no. B6, 10 June 1993, p. 9693. doi:10.1029/92jb01556.

Wootton, J. T., et al. "Effects of Disturbance on River Food Webs." *Science*, vol. 273, no. 5281, 13 Sept. 1996, pp. 1558–1561. doi:10.1126/science.273.5281.1558.

World Bank. "Part III: Cities' Contribution to Climate Change." *Cities and Climate Change: An Urgent Agenda*. World Bank, 2007, pp. 14–32, https://siteresources.worldbank.org/INTUWM/Resources/340232-1205330656272/4768406-1291309208465/PartIII.pdf.

Yeang, K. "Bases for Ecosystem Design." *Architectural Design*, vol. 42, 1972, pp. 434–436.

Yeang, K. "Bionics – the Use of Biological Analogies in Design." *Architectural Association Quarterly*, vol. 4, 1974, pp. 48–57.

Yeang, K. *Designing with Nature: The Ecological Basis for Architectural Design*. McGraw-Hill, 1995. Republished in Spanish by Gustavo Gili, 1999.

Yeang, K. *Ecodesign: Instruction Manual*. John Wiley & Sons, 2006.

Yurto⊠lu, N. "Plant Water Relationships and Evapotranspiration." *History Studies International Journal of History*, vol. 10, no. 7, 2018, pp. 241–264. doi:10.9737/hist.2018.658.

Zalasiewicz, J., et al. "The Geological Cycle of Plastics and Their Use as a Stratigraphic Indicator of the Anthropocene." *Anthropocene*, vol. 13, 2016, pp. 4–17. doi:10.1016/j.ancene.2016.01.002.

INDEX